'Businesses that are thriving in social media today do so by creating social capital with their customers – Jay makes it easy to understand why we must invest time and consideration in social before we can reap its rewards.'

Glen Gilmore
Social Media Strategist
Attorney and Rutgers Adjunct
Professor of Social Media Marketing

'Jay has been educating me about the workings of the social web on a daily basis for over 3 years via Twitter and Facebook. I am so excited to now be able to access all his knowledge in an organised book format. You too will gain great insight from this.'

Michael Q Todd
Social Media Strategist and Co-founder
of the agency Social Media Plus One

'Jay reveals how the world's social leaders gain business advantage from social media. He blends colourful business cases, artful analogies, and academic theory to create a passionate and practical handbook for anyone wanting to build their online reputation.'

Jeremy Woolf
Global Digital and Social Media Lead
Text 100

'The recipe for unlocking the true potential of social media–not just for marketing, but for leadership.'

Aaron Lee
Entrepreneur Blogger and Social Media Consultant

'A smart guide to digital social graces that will help you transition from informed bewilderment to enlightened pioneer'

Napoleon Biggs
Founder & Host, Web Wednesday Asia,
and SVP & Head of Digital
Fleishman-Hillard Communications Asia

'Your backstage pass to not only understanding but also effectively capitalizing on the power tools of influence at work in social media . . . a must read for anyone and everyone in business today!'

Debra Meiburg
Master of Wine, and Director
Meiburg Wine Media Ltd.

'Jay deftly interweaves a sociological analysis of why social media has become all-important during recent history with down-to-earth advice on how to become a Socialeader in today's world. A highly engaging and readable book – especially for naysayers of social media who have resisted participation in social media until now.'

Joanne Ooi
CEO
Plukka.com

MASTERING STORY, COMMUNITY AND INFLUENCE

How to Use Social Media to Become a Socialeader

Jay Oatway

A John Wiley & Sons, Ltd., Publication

This edition first published 2012

Registered office
John Wiley & Sons Ltd, The Atrium, Southern Gate, Chichester, West Sussex, PO19 8SQ, United Kingdom

For details of our global editorial offices, for customer services and for information about how to apply for permission to reuse the copyright material in this book please see our website at www.wiley.com.

Library of Congress Cataloging-in-Publication Data
Oatway, Jay.
Mastering story, community and influence : how to use social media to become a socialeader / Jay Oatway.
p. cm.
ISBN 978-1-119-94071-5 (hardback)
1. Social media. 2. Business communication. I. Title.
HM851.O34 2012
808'.06665–dc23

2012001797

A catalogue record for this book is available from the British Library.

ISBN 978-1-119-94071-5 (hbk) ISBN 978-1-119-94345-7 (ebk)
ISBN 978-1-119-94346-4 (ebk) ISBN 978-1-119-94347-1 (ebk)

Set in 11.5/15 pt Adobe Caslon Pro-Regular by Toppan Best-set Premedia Limited
Printed in Great Britain by TJ International Ltd, Padstow, Cornwall, UK

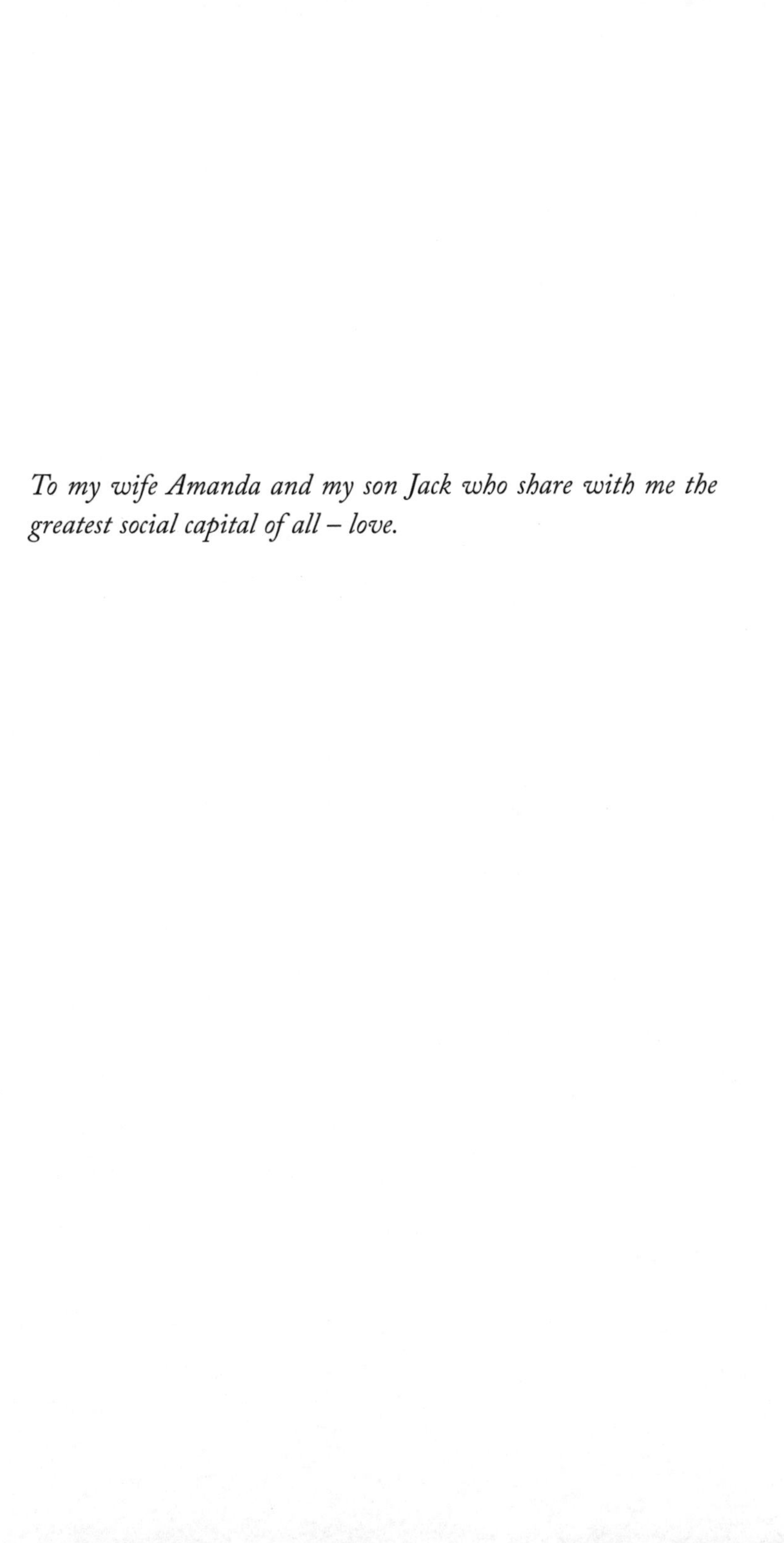

To my wife Amanda and my son Jack who share with me the greatest social capital of all – love.

CONTENTS

INTRODUCTION

There was a time not so long ago when a CEO would dictate his correspondence to his secretary. The notion that he type it himself would have been laughed at. Today, the modern CEO is on his Blackberry all day. Tomorrow, business leaders will be the masters of their own social media empires.

This shift has already begun. Increasingly, business needs people who treat social media as a professional thought-leadership tool, both for their own careers and for the benefit of the company they work for. We want to do business with those who make social media feel less like mass marketing and more like customer service. We seek out those whose influence has grown through caring for their community.

The future needs Socialeaders. A Socialeader is someone who treats social media as a professional thought-leadership tool, both for their own careers and for the benefit of the company they work for. It's someone who makes marketing feel more like customer service. It's someone whose influence has grown through caring for their community. A Social-eader acts as a role model in the workplace, demonstrating how to use social media tools professionally.

Why should you care about your digital presence, or how much online influence you have, or whether you know how to build social capital among relevant social media communities? Simply, you will be socially and economically disadvantaged if you don't.

An alternative economy is fast being built on top of social media. And those who ignore it, do so at their peril. Our digital presence tells a story of what we are worth.

At some point, you will be Googled, possibly by a prospective client, or by a new employer, or an investor. Over any matter of great importance, we will seek out more information on the person we are dealing with. We are being judged by what is found. But many of us have yet to try to close the gap between our offline reputations and our lack of online reputations.

Social media is clearly not a fad that will go away. We need to stop treating them like a child's toy and start using them like a tool of power.

Even if you haven't yet begun to take charge of your digital presence, there is already information about you online. You might not have placed it there, but anyone can find it. Why leave it to chance what people find out about you? Why not take the steps towards working with social media to curate an impressive living breathing biography of your expertise and reputation?

When that prospective client, or new employer, or potential investor does a search for you, and one for your competition, who will look most promising? Shouldn't it be you?

You can't escape the fact that social media is reshaping the competitive landscape. Business competition studies are showing that those using social media is gaining advantage. Your boss is going to want this too. At some point, you are going to be expected to know how to use social media for business, just as you are expected to know how to use email or the telephone.

Resistance to this will not benefit you. Already you are missing out on deals, discounts and other free stuff reserved only for those with significant online influence. What is online influence? Think of it as having what it takes to get into an exclusive club. All of our social media activities are being monitored to judge their relative impact within online communities. If you can't demonstrate that you have the influence it takes, then you can expect to wait a long time in the queue outside. Is that where you want to be? No, I didn't think so.

Socialeaders go straight to the VIP room. But you only get as good as you give. We need to begin to invest a significant amount of effort into mastering the new frontier. It's more about investing in the people than it is about the technology. It's not called social media for nothing. It has been said about this new technology that 'The last mile is human.'

Socialeaders are part of that solution; you are the human that completes the transition to the new way the world operates. You can't afford to be the last person to figure this out. The younger people in your company need leadership. They may be digital natives, but what they need is a digital role model to show them how to use social media as more than a toy.

Granted, it's not easy keeping up with all the rapid developments in social media. I use the term social media purposely throughout this book as a generic stand-in for the hot services of the day, like Twitter, Facebook, Google+, LinkedIn, YouTube or Instagram. I'm well aware that social networks come and go.

But I'm also aware that no matter how much the social media landscape evolves, the underlying principles at work only become stronger and stronger. That's what this book will explain so that, no matter what comes next, you understand the bigger picture of what needs to be done to be successful.

Social Currency, Social Capital and Social Credit in a Nutshell

There are three terms I use throughout this book to help explain what is happening in social media and how to use it to become a Socialeader. They are: social currency, social capital and social credit.

Essentially, story is the currency of social media. Through the exchange of social currency you build up relationships. The value of these relationships is your social capital.

Social capital works much like at a bank: it's hard to withdraw more than the value you have already deposited. You can't ask more from a relationship than you give to it, not if you want to maintain that relationship.

Also like a bank, the longer you invest in relationships, even if the social currency is of only modest value, the more social capital you will have.

The bottom line: to get more out of social media, you've got to put more in.

Social credit, on the other hand, is often offered even when you don't ask for it. It is born out of your online reputation (or, more likely, from a perception that you are someone important). It is when you are given the VIP treatment because your online presence is considered influential by someone who wishes to get on your good side. Social credit is a double-edged sword and must be handled carefully.

The Take Away Express

This book is about transforming you into a future-proofed Socialeader in a way that saves time and is easy to refer back to. To that end, you'll find a series of quick references to the main points, called *Take Away Express*, at the end of each chapter. These are designed to help make it easier to jog your memory about the many things Socialeaders need to remember.

Let's get you started.

Part One

STORY: TELL IT WELL

1. WHY WE MUST SHARE

'Those of you who think you are creating online content, take note: your success will be directly dependent on your ability to create excuses for people to talk to one another.'

– Douglas RushKoff[1]

Content isn't king, but the conversation is. It always has been. We need to be social with other people. Don't get me wrong; 'content' is important, but only because it gives us something to talk about, and a way to be social.

Many working in the business world today have never been required to be social with customers. Mass media marketing techniques took business people away from real conversations, replacing them with ad campaigns. But social media has brought those conversations back.

What we're realizing for the first time in decades is that we need to personally return to the fundamentals of passing along stories of value to the public, stories that come from our professional expertise but that can then be passed along again and again. We must learn to share if we want to have an online presence that matters.

We are all newcomers to this online realm. My background is in print magazines. I still have a deep love for magazines, even though I know I can get their content in digital form. I like the feeling, the experience. It's a bit like being a DJ who likes vinyl records. Someday these great bound collections of dead-tree pages and toxic inks will be consigned to a specialty store. Despite having absolutely no shortage of digital content, I still love seeking them out.

Those of us who come from the print era know that magazines and books are difficult to share. Sure, we can always give them to a friend when we are done with them, but then we no longer have them ourselves. Or we can recommend that a friend also buy and read them. Ultimately, our goal is to be able to sit and have a conversation about the subject matter which touched us.

Shifting to the web was an inevitable leap for me. For you, in whatever your industry, it will likely be inevitable too. The sooner you become a master of this new domain, the better off you will be. The web allows us to quickly share and consume interesting subject matter with other like-minded people, and then have conversations about those stories.

When I came to the web, I brought with me my magazine instincts. Curate: find cool stuff that I think my audience will like. Build up a big audience to share it with. Lather, rinse, repeat. Originally I thought, like many publishers still do, that the value I could monetize here would be the same that I monetized in the mass media era: the size of the audience.

But along the way I discovered something crucial, and now obvious: social media isn't mass media. This isn't broadcasting from on high. This isn't about putting ads in front of as many eyeballs as possible. This is something even better: creating enduring relationships with people who can unlock all sorts of new business opportunities.

This began with hanging out with my so-called readership every day – the same way I used to hang out with work colleagues around the water cooler. Albeit the water cooler was getting pretty big, it wasn't long before there were more than 10,000 – about the same as the monthly print run at the last magazine I worked for (I never knew a single one of my readers in the print days).

But my audience was also contributors. And the whole process clearly became a type of conversation – a sharing of ideas among a group of like-minded people.

These people are still there. So are the conversations. The relationships are deeper now. And the communities that we reach collectively are larger than ever. The trust we have within these communities is also strong.

But communities require constant attention. The more you can contribute the better. You need to be sharing stories, relevant stories, stories of value, stories that help the community make sense of the ever changing media landscape. You need to be sharing about that which you have expertise in. Or that which you care about. Or any ember of passion you feel towards anything. Don't worry, if you're not sure what this is, I'll help you cultivate that later in this book.

For now, let's just make sure we are clear on the importance of sharing in this decentralized network era. Sharing is the foundation of conversation, the foundation of social media and the foundation for all relationships.

There is a lot of psychology at work in a relationship, and therefore to be Socialeaders we will need to develop an understanding of why we 'like' different things. This will help us pick and choose the most valuable content to share.

You Know How to Share Gossip

Professor Robin Ian MacDonald Dunbar, British anthropologist and evolutionary psychologist, never set out to explain social media, but much of his work has helped Socialeaders make sense of what is going on today.

Dunbar argued that gossip, in its broadest sense, is a 'fundamental prerequisite of the human condition'.[2] From my experience in social media, I'd have to agree. So unless you are somehow living outside the human condition, you can't really say you don't know what gossip is – or how to partake in it. This means you understand the basics of what makes social media tick.

The content that we share online is sometimes referred to as social currency. It's a useful way of thinking about content. Sometimes it has more value, sometimes it has less. It depends on the context of the situation and who is involved in the conversation. I'll go more into that later, but for now anything that feels like gossip to you will likely make for turbo-charged social currency:

> *'What is told in the ear of a man is often heard 100 miles away.'*
>
> **– Chinese proverb**

> *'Do not repeat anything you will not sign your name to.'*
>
> **– Author unknown**

The last anonymous quote, quite ironically, is crucial. Everything we share through social media has our name, or our company name, on it. So we must be careful what we share online. This isn't 'off the record' gossip in the back of a bar.

Social media is public record, and no one wants a reputation of being a horrible gossip.

Gossip, even in the broadest sense, has had such a bad rap for so long that we can't really talk about it as a useful tool. So let's call it 'story sharing'.

These stories can be false, faked or otherwise untrue. Stories can misrepresent and take matters out of context. Stories can be cold hard truths or complete works of fiction. But more often the stories are 'true enough' and they matter. They have to be trustworthy, come from a trusted source and carry information we can act on because without story sharing, we can't sustain communities. Our story sharing acts like building maintenance for societal groups. Without it, our groups crumble into rubble.

Dunbar asserts that our story sharing 'is the central plank on which human sociality is founded'.[3] It is certainly the central plank on which social media communities are founded. And it will be the plank on which we build our own online presence.

The stories you share will determine the crowds you attract, which will determine your online influence.

So you need to begin with the end in mind. The central plank of stories you share from the outset will determine where you end up – because it will determine whom you begin to bond with.

Sharing Stories Creates Bonds

Try getting through a whole day without sharing some sort of 'news' with someone. You would be a pretty lonely person. It's less about the content and more about the social bonding.

But bonding is not what we first think about in the rush to 'engage' in online conversation. Instead of asking, 'What exactly are we going to engage customers about? What was this so-called conversation supposed to discuss?', the tendency is to start pushing out brand message the way we were accustomed to doing in the Broadcast Age.

Too often, like some self-centred brute on a first date with an attractive woman, we meet with our customers (virtually) and proceed to talk non-stop about ourselves. It doesn't take long before they tune us out and start looking at their watches. That is *not* bonding. That is not what is meant by sharing, engaging or conversations.

Broadcasting isn't really sharing. Putting out your own carefully crafted message serves your own purposes. It serves your brand, your product, your service. But rarely has this served others very well. Seldom have we provided a story that, like gossip, can continue to be shared and, every time it does, deepen bonds between people.

The catchphrase I hear all the time among serious social media practitioners is 'sharing is caring'. While this platitude sounds like something your kindergarten teacher might have told you, it is actually an important sociological type of 'social grooming'.

Social media is more about the 'social' than they are about the 'media'. To become masters of this medium, we need to think like monkeys. Yes, monkeys. Caring monkeys that groom each other. Sharing stories of interest among our like-minded monkey tribe is a bit like how real monkeys pick fleas out of each other's fur. This lights up our primate brains with an emotional charge and is an age-old, fundamental bonding ritual.[4]

But the stories you share must be valuable in some way to those around you. You only bond with fellow monkeys who are good at picking out your fleas. And to be good at picking out someone's fleas, you need to know where they scratch the most.

To know where someone scratches requires connecting with that individual one-to-one. It's about you, not your company – YOU! To be successful with social media, you need to keep in mind that you are going to be living with yourself online forever more. Rebooting your online presence years from now will be nearly impossible. We need to take responsibility for building relationships, and that begins with taking responsibility for finding great social currency to share.

What is Social Currency?

It's important to understand that there is something a little bit more complex about the idea of sharing stories to build relationships. This is where the term 'social currency' comes into play.

The term, as I'm using it, was coined in 2000 by the American media theorist Douglas Rushkoff. Social currency isn't a new term to those working in the world's top advertising agencies or to those who study sociology, where the term has been in use for much longer. It gets talked about a lot these days. But like everything else that gets talked about a lot, people often get it wrong.

To be clear: social currency is the information you acquire, then share, to start, maintain and nurture your relationships. It's important to think of it as a bit like money, although not as a direct analogue. If you give social currency of value to someone, you can then pass it on as something valuable to another person. With real money, when you give it away, you have it no more. However, when we share stories, we still retain those stories, and it's the excitement of the conversation they evoke that we truly value.

Viral videos, hot gossip, good jokes: these are social currency with a high pass-on value because they get us talking. When we are sharing stories to create bonds with other like-minded people, we want to give them social currency with the highest pass-on value we can. The more they can share what we give them, the more people they can talk about the experience with, the deeper our relationships become.

Social currency can help us feel like we belong to a community, that we get an inside joke, or otherwise feel significant.

If this is the first time you've heard the term 'social currency' expect that it won't be the last. Often when we finally have a term like 'social currency' to describe what we have seen going on, we begin to see it in action all around us every day, and everywhere we go. You will see social currency exchanges every time you see people having conversations.

Social currency forms the 'central plank'. Sharing it is the 'social grooming' that holds communities together.

How We Lost Our Way and Have to Find it Again

Social currency exchange has been with us since the very beginning. The reason it feels so new in the social media age has a lot to do with where we've been during the past half century or so. We need to put the 'media' side of social media into perspective so we can better make sense of the revolution we are living through.

Technology has, many times, altered the course of society, by making it easier to exchange social currency. The arrival of the printing press marked a dramatic shift away from what had been mostly a one-to-one exchange. Newspapers ushered in the dawn of mass media.

Newspapers provided us with our daily 'social currency', the news and opinion that we traded with colleagues, friends and family. Everything from politics to sport: more than enough to sustain you through all those encounters with other people during the day. No one bought the paper so they could read it and keep the information to themselves. Newspapers were designed to be discussed and argued over. It was all about the conversations, just as it is today.

Then came radio and television. And the Publishing Era begot the Broadcasting Era. Mass media got even more massive. Millions of us were all reading and watching (and talking about) the same things, selected by relatively few people.

Let's remember that this was a radical departure from anything that had gone before it. Mass media was a powerful stage, with high barrier to entry. Its reach and influence were beyond anything previously possible.

This made mass media ideal tools for promoting brands, and with their arrival began a tradition of businesses talking *at* consumers, indirectly, via advertising. Ads were also big business and created a business culture to which the concepts of 'conversation' and 'engaging' meant running focus groups.

Soon advertising became all about the art (and science) of persuasion. And while ad men analyzed audiences for demographics (age, sex, income bracket), they treated everyone in that demographic as equal in terms of their respective influence (or lack thereof). There was no way to identify or create a special relationship with the ones who were particularly well regarded around the water cooler.

With the rise of mass media and mass marketing also came mass production. Retail began to evolve to take advantage of this scale. With this came the demise of something we all seem to cherish: the small-shop feeling of being a customer known by name, treated with individual care and respect. We lost our sense of community.

I was born into this era. I watched my small town turn into a much bigger town and the local hardware store, convenience store, grocery store – even the independent video rental store – all disappear with the arrival of the big box super stores.

Now I'm not here to disparage the big stores. I just want to illuminate how we've reached this cross roads. The mass media era wasn't the default state of human civilization, but rather a brief time of exception – a time that, with the emergence of social media, is coming to an end.

This, more than all the psychological reasons, is why we must share. Why we must cast off our mass media thinking and become Socialeaders. There is a massive change coming to the way we do business. The rising 'social' component requires a return to the mom and pop shop mentality. It requires a 'humanization of business'.[5]

If we want to see the canary in this coal mine, we need only look to the newspaper that ushered in the mass media era. We don't depend upon the newspaper for our social currency. We can find our current events, our reviews, our sports, our coupons in many more relevant places online. In fact, these things often find us now through social media. As a result, circulation rates for the vast majority of newspapers and periodicals have been on a steady decline.

And it's not just the newspapers. All mass media have been impacted by the wave of change that is sweeping across our societies. It's no secret that we aren't buying music like we used to. We aren't watching TV like we used to either. We aren't shopping like we used to. We aren't behaving at all like we used to during that brief era of mass media and mass marketing.

We *are* behaving like we used to behave before that era.

Social media has brought us back. And for those of us who can't remember how it really worked before, we need to seriously start to understand how to connect with people again.

As the newspaper did before, social media acts like our central bank of social currency. It is online that we will find everything we need to talk about. And blogs are the first stop.

Blogs began as ego projects, and bloggers were dismissed as people with too much time on their hands. Today we can no longer afford to treat blogs, Facebook and Twitter as toys of ego. That 'egosystem'[6] has evolved into the professional production centre of value for our social currency economy.

Blog posts are especially important because, in a radical departure from mass media, they are easy to share. In fact, most bloggers want you to share their social currency – not outright steal it, of course. But if you are willing to give them a mention and a link (the standard for respectable, social capital building sharing), then you are welcome to turn it into social currency to share with your community.

With blogs and later the micro-blogging tools, our personal oral tradition went digital, and the exchange of valuable social currency accelerated at a breakneck pace. Every water cooler and bar room conversation seemed to suddenly be online.

And here is where the new problems begin. Since both blogging and sharing of blog posts through social media is easy to do, the noise levels have grown and grown.

Filtering the noise to find quality stories and share them effectively among all the noise became, and remains, the valuable craft. It's this craft that attracts an influential and loyal audience. It's a craft that used to belong to news editors. It's now the turf of Socialeaders.

Take Away Express #1

- ☐ Conversation is king. It's the sharing of content and the discussions that ensue that truly matter.
- ☐ Everyone knows how to share gossip to maintain social bonds. So no one can claim they don't get how to use social media.
- ☐ Prepare to interact with people one-on-one. Social media isn't mass marketing.
- ☐ Story is the currency of social media. Building relationships requires exchanging social currency.
- ☐ Social media is like a bank. You need to deposit before you can withdraw.

2. A BETTER STORY OF YOU

'We construct a narrative for ourselves, and that's the thread that we follow from one day to the next. People who disintegrate as personalities are the ones who lose that thread.'

– Paul Benjamin Auster[1]

In the social media age, you are now responsible for writing your own history. There was a time when that job was the responsibility of historians. But, as the number of historians is few, and the numbers of the rest of us quite large, history managed to forget about a lot of really cool people. This won't be the case in the future.

What you are creating online today is a living, breathing curriculum vitae: a history of yourself that goes beyond a mere résumé of your career to expose a much richer and more detailed 'course of your life'. This is your tapestry of likes and dislikes, of thoughts, ideas and opinions. Social media records everything we are, as we express it and share it. Every exchange of social currency becomes part of your autobiography.

So, as privacy advocates will remind us, you need to be mindful of what you put online. You need to be clear about what you want to be known for and where your expertise lies.

The good news is that you are in control of the story: as the editor, the actor, the narrator. You are also the marketing person building relationships with your audience. This makes you the most important part of your story. No one else can take responsibility for this. The future wants histories told by the actual people who live them.

Up until now, you've probably intentionally been keeping a bit of a low profile online. Why? The reasons are often concerns over privacy, or fears of reprisals, and sometimes it's just apathy.

Whatever the reasons you might have had to stay off the radar, you need to set those aside and say to yourself: 'I am the master of my own show.' Then you need to get down to the business of showing the world just who you are.

The alternatives aren't pretty, especially when Googled. For those who aren't actively sharing social currency online, one of two results return for searches on you. The first is that you don't show up at all. While some of you may breathe a sigh of relief at that notion, you've got to think about your competitiveness going forward. At some point you are likely to be Googled as part of an assessment for a new job or promotion, or by a client deciding whether they should choose you or a competitor. Having no results is not going to help you when the other candidates have pages and pages of references to their thought-leadership.

The other scenario is even worse. At some point you may have some sort of embarrassing moment immortalized online. If there is very little else about you online, then the embarrassing content will persist at the top of all searches. Those with large online profiles tend to return search results to a lot more social media than they control, burying the embarrassing stuff under a mountain of material they can be proud of.

Your history is in your hands.

Accept it: You Are Now an Authority

Don't be shy. It's perfectly acceptable for you to identify yourself as an expert in your online profiles. In fact, it goes a long way towards attracting a following. The more easily people can recognize your qualifications, the more likely they are to listen to you.

That doesn't mean boasting about your credentials all the time. Instead, craft a clear social media bio that you can use on a wide range of sites. Have a short, 160 character version and a longer, two-paragraph version. Include in it your position, your passion and any status that may be relevant.

Then you need to dress the part. You profile pages should include the most professional picture of you possible. That doesn't mean the picture they used for your employee pass. Have a friend help you shoot some 'professional' images that show character. If you can afford it, hire a photographer to do the shoot. The quality will scream 'professional'. Remember, this is an investment in you.

But it's not all style. Substance matters too, including your unique perspective, and your ability to choose the relevant stories to share. This is what will make you the interesting 'go-to-guy'. It's what is going to attract qualified leads (and save you from having to hunt them down).

Authority is also a product of building up trust through making people's lives better. Authority, while I say you should state it yourself, truly comes from the community of people you've helped. It's part of reputation. We'll look at how this works in Part Three of this book, but for now I want you to stand proud and not be afraid to identify yourself as an authority.

Turning Your Personal Accounts Professional

You absolutely *must* have your own personal social media accounts. After all, you are going to be stuck with you for the rest of your life.

In the early days of the egosystem, we played with social media as a toy, which is understandable. It's initial appeal wasn't for business; it was for having a laugh. It was about connecting with a few friends. This is also partly why social media was so quickly banned from so many workplaces.

But as social media began to mature, so did its users. Many early adopters of social media began to experience a blurring between their 'personal' social media, which they've tried so hard to keep private, and their 'professional' social media, which they've tried to keep their own faces off of.

It goes to show how tricky it can be to have work and personal so close together. This is the new norm, and we need to learn our lessons from the mistakes of those who have bravely gone before us. These are no longer uncharted waters. The clear leaders are those who can go pro, and yet gracefully include a bit of themselves in the story.

It's time to clean up your 'personal' accounts and come out of the closet if you can. It's time to dress for success and show the world the professional, public you. This means relegating the stuff for close friends to places that only close friends can see. And, even then, being very careful what you share, knowing that once you put something online, there is a good chance it can escape, irretrievably, into the wild for all to see.

Concentrate instead on sharing a lot more thought-leadership links. These should be the bread and butter of you daily social media diet. Adjust you profile settings so that nearly everything you share is open for the world to see – and to search.

You are a public exhibit in the world's biggest fair. Make yourself interesting, unusual, bring out twists, make it look sharp, smart, quirky, or whatever it is that you value. Cultivate a delicate balance of professionalism mixed with personality.

Not Told All at Once, but Gradually through Choices

The story that fills the screen of your living breathing CV isn't actually about you. At least not directly. It's about your choices. Social media is not about creating brochure-ware – pretty webpages that only get updated annually. We don't draft our autobiography and then place it on a static page. Not as long as we are alive. Instead, the story unfolds with each day, each choice, each social currency exchange. Socialeaders don't just tell, they show. For example, if your professional story was about 'holiday' then you wouldn't wait until you got home to show an unedited slide show of every single picture of your trip. You would share the experience as it unfolded. It would involve little check-ins at cool landmarks. It would feature exchanges with locals about the best places to eat. It would showcase the wonders of getting lost and discovering hidden treasures. Bit by bit, the adventure would play out in real time for everyone to share in from afar.

For most of us, day-to-day life isn't a holiday, but something much more grounded in the workplace. But it's from our computers that we do follow along from afar the news of our interests, business and industry. It's the news about these things that we begin to curate as part of our own story.

Each day, we make choices as to what is hot or not. Then we click the share button, and by doing so demonstrate our expertise. We reveal ourselves not by what we say about ourselves, but through our expert judgement as to what matters, and what is worth sharing.

You will soon find that you are better than you think at knowing what matters in your area of expertise, just as a sommelier knows how to select the right wine. And for this people will trust you.

At a conference once, I'd made a recommendation for a new social media tool that I thought people in the room might want to consider. After I spoke I had a fan approach me and proclaim, 'You say sign up, I sign up.' I was more than a bit concerned by their blind faith. But I've since realized that that is the way trust works. You don't earn it all at once, but gradually. And you must always guard it. So take care in the choices you make: they may soon be influencing the choices that others make.

No One Quite Like You in Your Master Narrative

There is a term I want to introduce now: 'master narrative'. The idea is that we are each part of not only our own stories, but also of a much bigger story. This bigger story includes all that is going on in and around your business, your industry, your geographic location, your politics, your religion, your science, your sports, your lifestyle choice and more. This is the master narrative that we live in, with many, many somewhat similar other people, all creating things, consuming things, exchanging social currency, forging bonds and exerting influence. Your master narrative isn't just your niche; it's also your new 'mass market' that you will look to gain a share of. Think about how you are going to build a share of your master narrative.

There are many master narratives and I'll explain later in this book how to find yours. For now let me use the master narrative that I live in as an example so I can illustrate how to stand out from the crowd and add value to the social currency of the day.

I'm a geek. I love technology. I share technology stories with other geeks every day. Some days, we are all talking about the same story, like when Apple announces a new iPhone. There is so much excitement on days like those that it feels like Christmas for geeks.

Everyone is sharing every scrap of news they can. But that sort of social currency isn't highly valued, since there is so much of it circulating on those days. So what can I do to stand out?

I insert a little bit of me. That's not to say I make the story about me, that would be a mistake. But I do try to inject my own sense of humour, my own thoughts and opinions about the 'news' of the day. This is still professional – but it is also personal.

No matter what field you are in, no matter how crowded your market, you are different from everyone else. People may disagree. They may laugh. But that's what you are hoping for: an emotional connection. When you get that, then you are mastering not just social media, but your master narrative.

The Currency of Your Story

As you start sharing stories via social media, you'll notice that certain stories get a lot more reaction than others. While there is no sure fire way to know what others in your community are going to pick up on, you can be sure that stories with an emotional charge will make the most valuable social currency.[2]

Heard any good jokes lately? That's a question that has been with us for a long time. And when a friend replies, 'Yeah, I've got one for you,' he immediately grabs our attention.

Of course, the joke isn't always funny. That's why those who know how to tell good jokes are so valued. Jokes are essentially little stories with a twist or surprise that evoke an emotional response.

The same is true for stories you share via tweets and Facebook posts. They don't have to be jokes, just the news of the day. While not every story comes with emotional charge (often news in your business may be important but a bit dull), try to add a little of your own wit and charm to everything you share. It won't hurt. If you can take a headline that's either already humorous, or make it a bit funnier, it will become powerful social currency when you share it. Don't underestimate the power of making people laugh.

But funny isn't the only emotional response. You can also try to tap outrage. When you come across an injustice, be sure to share it. If it is something that makes us want to shout, we will pass it on as we add our own voices of disapproval.

It works for the whole spectrum of emotion: anger, shock, sadness, sympathy. But know that it's harder to build a community around a lot of anger than it is to build it around a lot of humour. Some of the most influential people on social media are comedians.

Beyond emotional charge there are other factors that you need to consider to determine the value of your social currency. Most important is timeliness. Serving up old news, unless you have something new to add to it, won't attract a lot of interaction. People will value you for your thought-leadership if you are fresh and up to the minute. They will tune in to you because you are a source of important new developments that they don't want to miss out on.

The exception to the timeliness rule is your 'back catalogue'. The longer you are sharing and creating content online, the more information builds up in your archives. This back catalogue is searchable. Google will, on a daily basis, send people to blog posts from years ago. It's not uncommon to find that a post from a couple years ago suddenly gets a spike of views as someone tweets it for its value, even though it has little timeliness.

That's a great opportunity to turn something old into something new again. Recognize that this is 'evergreen' content. Do a follow up. See if you can find who was sharing your old postings. Reach out and thank them. Offer further insights.

But, as always, the real test for the value of social currency is how much your audience feels they stand to gain by passing it on. If your social currency can become their social currency, and in turn get their audiences excited, talking and sharing, then you will find you have a community that keeps coming back to you for more.

Until you start to get a feel for what makes for valuable social currency, just ask yourself a few questions about everything you are going to share:

- Will it help people feel like they belong to my community?
- Will they feel an emotional charge?
- Will it answer a question or solve a problem?

If the answer is yes, then share. If no, can you tweak it so it does? Just be careful not to tweak it so much that it becomes counterfeit social currency.

Counterfeit Social Currency

Bald face lies often travel surprisingly well. We've seen many hoaxes sweep across the social media landscape. Mark Twain once wrote: 'The reports of my death have been greatly exaggerated.' Those exaggerations happen a lot through social media. It's important not to get entangled in them.

We need to think of the 'lies' as a sort of counterfeit social currency. They're designed to evoke an emotional response, which in turn gets them passed on. But they damage not only those whom the lies concern, but also those who pass them along.

Death hoaxes continue to plague celebrities. Those who've been incorrectly pronounced dead via social media include: Paris Hilton (2007), Miley Cyrus (2008), Mick Jagger (2009), Kanye West (2009), Johnny Depp (2010), Justin Bieber (2011). Natalie Portman, Tom Cruise, Tom Hanks and Jeff Goldblum were all reported to have fallen off a cliff while rock climbing. No, not together at the same time. Separately, but all falsely. Which, incidentally, shows how certain ideas like falling off a cliff seem to make for better counterfeit currency than perhaps getting hit by a bus.

Remember, people don't like to be deceived, but mistakes can happen. We all can get swept away by our emotions. In this real-time media environment, it's far too easy to make a response before we have checked all the facts. It is essential that when this happens we correct the mistake as quickly and prominently as possible. Apologize to everyone. And make sure to put our misinforming sources on notice that we won't accept this sort of currency again. If this source has burned us before, then shame on us. Never again.

We all need to think more like journalists. We need to do some fact checking before we share anything that sounds too, well, unbelievable . . . it likely is. I make it standard practice to open and verify all links to stories before I pass them on. If the link leads to a story that links to another source story, I click through that link and check it out. I keep following the links until I reach a source that I trust (usually a news site or well-established blog).

But sometimes you have to know when *not* to click a link. There's a term for this: it's called 'link baiting' and it's despicable. Link baiting prays on your emotions, often saying things like 'I can't believe what you said about me in the post' or 'I can't stop laughing at you in this picture'. Our instinct is to immediately react. The emotion response is high. But clicking on this could spell disaster. This is one of the most common ploys for spreading viruses. You need to train yourself to remain calm when you see these sorts of things coming through your social media channels, even if they are coming from trusted sources. They possibly fell for the trap and now one of their accounts has been compromised and is spreading the virus forward. The same thing has happened before in the world of email.

Link baiting is also the foundation for the Rickroll prank, which can serve as a useful training tool in how to treat links that sound too good to be true with a heightened level of suspicion. It can also be a lesson in how to get a bad reputation through sharing counterfeit currency – after a while, no matter how much you produce, no one is going to accept anything from you.[3]

When Rickrolling first began, it was 'funny' currency with just enough positive emotional charge that actually worked in a positive relationship-building fashion among certain early adopter communities, circa 2007. It was much like schoolboys playing pranks on each other.

The Rickroll joke goes like this: an email or tweet appears, usually from someone you trust. It reads something like: 'The first trailer for [insert highly anticipated cinematic event of the year]' and then it's the basic 'bait and switch' con. The link, which is shortened to disguise its true destination, sends you to the music video for the 1987 Rick Astley song 'Never Gonna Give You Up'. When a person clicks on the link and is led to the video, he or she is said to have been 'Rickrolled'.

I've been Rickrolled. Twice. Fool me once, shame on you. Fool me twice, shame on me.

This isn't 2007. Today, it has to be about sharing quality links on Twitter and Facebook, not springing practical jokes. There is too much malware floating around already to allow our reputations to be tainted. Likewise, don't pass on link bait by accident. Be sure you check out everything you share, while not falling for link bait leading to malware.

Trust is precious: it's hard to earn, and easy to lose. Don't exaggerate claims to links you are sharing. Manage expectations. Misleading people even in the slightest breeds only disappointment and distrust.

Develop trusted sources. Be suspicious of anything that comes out of the blue from someone you don't know. Check out the credibility of those who have passed it to you. Use a tool like Tweetdeck to expand any shortened URLs that you don't know where they lead. And upon the news of a celebrity death, always, always get confirmation from a second more credible source. If it sounds too good to be true, it's likely Rick Astley.

Photos, Music and Video as Social Currency

What I've been describing thus far in terms of social currency amounts to basically the headlines and links to news stories. I think this makes for a strong foundation to any thought-leader, but with so much competition for people's attention, there is a need for more pizzazz.

This gets tricky though. When Facebook introduced the new Timelines in 2011, one of the first apps people were using to automatically share social currency in their living, breathing autobiographies was Spotify, the music application. Spotify revealed something that many of us long suspected: in the absence of mass media radio, we've all become our own DJs. Which is fine, since we only have to please ourselves with what we choose to play. But the Spotify app, by default, immediately began telling all our friends what we were listening to. This led many users to cry 'You all have lousy taste in music!' To which dozens replied, 'So do you!'

Obviously, most of us aren't music authorities. We don't claim to be. Those who are hang out on social media sites like turntable.fm, where their superior ability to choose crowd pleasing hits has garnered them large audiences.

Music can have a powerful emotional charge. While it's tricky to know how to please everyone, sometimes, when we know our audience well, it is possible to pick something that pleases a few people. Try using sites like Blip.fm,

Soundcloud or even YouTube to share some music that punctuates a conversation. Often people are really touched by this.

Another way to really touch people is through photography. A great picture shared through photo sites like Flickr or Instagram can be a powerful way to sum up the mood of an event, or relate an experience. There's a reason we say that a picture is worth a thousand words. Every phone now has a camera. Use it to your advantage. Seize opportunities to make unique social currency. Never fail to have your picture taken with anyone you deem an authority in your master narrative.

The most powerful form of social currency is also probably the hardest form of all to do well: video. Sharing clips isn't such a problem, but creating your own requires the greatest amount of production skills of all the media. But that's also what makes it such a powerful tool for conveying your thought-leadership.

If you are going to use video, remember that production value is everything. Make sure you look and sound good. Content is almost secondary to that.

It's possible to grab passable video using your camera, but reserve this for those spontaneous opportunities like when you get a couple of minutes alone in the wings of a conference with a major influencer. Get them to answer a couple questions relevant to the conference and immediately post it to YouTube so your followers can get a better sense of what it's like to be there.

But not every day is about big events and meeting celebrities. It's important not to misjudge the value of small talk. This is the oft-maligned 'what I ate for breakfast' content that naysayers claim is the bane of social media. I disagree. Small talk is the glue of relationships. When we chat with our closest friends, we seldom speak exclusively of the biggest news of the day. We ask after each other, 'How've you been?' Or we talk about the weather, 'When will it stop raining?' These little exchanges between people are called 'pleasantries' because they are, in fact, pleasant. 'Pleasant' doesn't sound like super hot social currency, but over time it goes a long way to building social capital.

The final, 'light content' social currency is quotes and motivational expressions. These don't suit everyone's thought-leadership, but if you can work them into yours, you'll find they have a lot more emotional charge and therefore evoke a lot more response. For light content, quotes can pack a big punch.

At any rate, look beyond using headlines and links all the time and you will get a lot more attention.

Currency Conversion

There is a hierarchy of social currency.[4] At the bottom is data: raw but powerful in the hands of someone who can interpret it, which leads to information. Information always needs to be verified and combined with other verified information before it can be converted into the next level of currency: knowledge. Knowledge when mixed with creativity can spark the next level: innovation. Innovation when held in perspective by those with experience becomes the most valuable currency: wisdom.

Performing currency conversions goes beyond the practice of mere curation, and adds a lot more original value each time you transform something simple into a higher form. The more you can work on generating innovation and wisdom from the raw materials that you find on the web, the greater your authority will be.

Easier said than done. When we find stories in our daily reading that we want to share, that's not the time to be doing currency conversion. We just carry on sharing. But there needs to come a point, towards the end of the week or month, when we reflect back on what's been going on, what's been important and try to pull some of it together into a more valuable form.

Infographics are an example of how people and businesses are doing this. Infographics create a lot of value going from data to information to knowledge, which not only make them hot social currency, but demonstrates the expertise of the creators. Infographics, even if you don't make them yourself, are always strong social currency because of the conversion that has taken place.

There is also a need to produce value at a consistent rate for it to have a positive effect on your authority and reputation. To be brilliant once doesn't seem to be enough. The lift from that will fade over time. So, the greater the frequency with which you can share your curation, creation and conversion, the more it will boost your influence.

To Maintain a Secret Identity or Not

The VP in charge of Google Plus, Bradley Horowitz, has suggested that there are three levels of identity: identified, anonymous, pseudonymous.[5] I like to think of those levels of identity as your real life, your private life and your secret life.

For the most part, you need to be the real you at all times in social media. The autobiography must be yours. Your future depends on it being valuable.

Then there is your private life. This needs to be kept off the radar. Seriously. Just keep it private. Which really means: don't share it online. It's a simple as that.

And that brings us to the third class of identity: the secret identity. This requires a pseudonym. I like to think of it as a bit like a Clark Kent and Superman sort of thing. Many people feel the need for a secret identity, a way to still share in the stories that matter to them, to engage in a master narrative that they love, but one that their family, friends and co-workers might take a dim view of.

Leading a secret life, I'm told, is not easy. But for many, it is the only way they can truly be who they really are. This means they need to create a fake bio, a fake picture, a fake name, and establish relationships with those who have no contact whatsoever with their 'other' life. Of course, the social media suites today make it easy to manage multiple identities if you want. But be careful: it is very easy accidentally to send a message out under the wrong identity. While these tools can save you time having to log in and out from various identities, they can also inadvertently out you.

If you aren't the sort of person who is 'in the closet', then don't hide behind fake names and an unidentifiable profile picture. At the time of writing, Google+ is insisting on real names. Facebook has had a similar policy. And I tend to agree. Blog commenting systems where real names are used (and, even better, when hooked into real identities on big social media sites) tend to breed much better discussion.

But there are two other types of pseudonymous identity: the performance artists and the branded account. Performance artists include accounts like Darth Vader, God and a whole army of angry Hulks. They are all quite funny, making light of the current events of the day. As I mentioned, humour makes for valuable social currency.

The second type is the branded account. These are accounts owned by companies and operated by a nameless, faceless individual. There is someone behind it, but who knows who that is. Without a true name or face these are difficult to trust. Which makes me wonder why so many brands choose to create this sort of online presence. Who are they hiding from?

Take Away Express #2

- ☐ You must have your own accounts. This is your living, breathing CV. You are writing your own history with everything that you share.
- ☐ Make your account professional and public. Identify yourself as an authority, not just through words, but also appearance.
- ☐ Tell your story carefully. Make wise choices and recommendations.

- ☐ Use your individuality to your advantage. Think in terms of capturing a bigger share of your master narrative.
- ☐ Emotional charge makes social currency valuable. Add in your own charge. Mix it up with music, photos, videos and quotes.
- ☐ Beware of counterfeit social currency. Develop trusted sources. Check everything before you pass it on.
- ☐ Don't hide behind a secret identity unless you absolutely have to.

3. SOCIAL MEDIA IS LIKE A COCKTAIL PARTY

'We have two ears and one mouth so we may listen more and talk the less.'

– Epictetus, ancient Greek philosopher

Imagine yourself at a business cocktail party, one of those industry networking affairs. You are standing there having an interesting conversation about a topic that you and a few industry acquaintances share a common interest in. Imagine someone who just arrived at the party bursting straight into the middle of your conversation just so they could tell you about all the fabulous deals their business is offering that week.

You would look at the interloper as if he were deranged. Yet, that is precisely how so many brands have thus far treated the social media party. Instead of first stopping to listen to what everyone else is talking about, they simply start broadcasting their own message.

Opportunities arise in many forms: sometimes when someone asks a question; other times when they are sharing a story we know more details about.

Whatever form the conversation takes, listen for ways to make people feel more a part of the community. Listen for opportunities to respond with jokes, tips or anything that has the feeling of gossip (again, in the broadest sense). Offer up anything relevant that can keep the buzz of the party going, and ingratiate yourself as a valuable member of the conversation.

Sometimes the smallest of exchanges of helpful advice can go the furthest.

Monitor More than Your Brand

While there are many companies out there selling advanced social media monitoring solutions, many of the best listening tools in social media can be very simple.

Gary Vaynerchuck is a master of one of the simplest tools: Twitter search. Vaynerchuck took over his parents' liquor store and transformed it into a large-scale wine retail store called Wine Library. He then started a video blog where he would talk about wine in a down-to-earth manner. Vaynerchuck's story is about being a man of the people, rather than a wine snob.

And he lived that story by venturing out into social media and really listening to the wine conversations of everyday people. Vaynerchuck claims to have spent 12 hours a day using Twitter search. But he wasn't searching for the word Vaynerchuck, nor even his Twitter handle @GaryVee. He wasn't searching something as broad as 'wine' or mentions of Wine Library. Instead, he would search for all the mentions of very specific keywords, like varietals and wine regions. He would then help answer people's questions about wine.

He would help people: no strings attached. He didn't sell them on wine from his store, he simply provided a free service. This goodwill grew his wine business for him. It acted as social currency, and proved much more potent than any advertising. He claims his wine business is now worth more than US$60 million.

I, too, track all sorts of insider terms. For example, I watch for expressions like 'crackberry' when I want to know what mobile phone geeks are thinking about Blackberry devices. Seek out the insider lingo, jargon and acronyms that only other passionate experts would use.

From simple tools like searching within Twitter and Facebook using their own built-in search tools to more sophisticated social media monitoring platforms like Radian6 or PeopleBrowsr, at the end of the day the results are only as good as the operator.

Don't be content with just tracking your brand or product mentions. Dig deeper into all the terms that matter to your master narrative. Use your social media suite to create columns that allow you to quickly take the pulse, or even to drill deeper.

Use the monitoring to find opportunities to interact with someone. Be ready to share valuable social currency with them or, even better, to bring them into the story in a more meaningful way.

Listening for Opportunities

Unlike traditional media campaigns that are set up months in advance and run for a set period of time, social media creates brand opportunities on the spur of the moment.

During the 2010 Ashes cricket tournament, a young American woman, Ashley Kerekes, with no knowledge of the sport found herself, like a modern day Alice in Wonderland, transported down a rabbit hole to another world – all because her Twitter username just happened to be @theashes.

'I AM NOT A FREAKING CRICKET MATCH!' was her initial response to the sudden inundation of tweets as the cricket test drew near. Her tweets were usually about knitting and how to take care of children, but then the magic of Twitter began. Fans started a campaign: #gettheashestotheashes.

Brands like Qantas and Vodafone, to their credit, were quick to seize the opportunity. They had been listening to what was going on. And the timing was perfect for Qantas.

Qantas had been dealing with a PR nightmare after many of their engines were found to be in need of repairs. Even with the problem resolved, they were looking at starting a 'campaign' to get customers trusting Qantas again. So when they heard via Twitter that cricket fans wanted an American girl flown to Australia, they jumped at the opportunity. They flew Ashley from New York to Sydney, while Vodafone, also no stranger to listening to social media, got her tickets to the matches and a phone to tweet her experience.

From that point on @theashes was an instant celebrity, tweeting about the whirlwind adventure she had suddenly found herself on. Her journey had her brushing shoulders with cricket royalty, both players and media. All the while, she offered a fresh perspective of the game that delighted both hardcore fans and players alike. For the Australian media, which had little good news to report about the matches, @theashes became a feel good story about the hospitality of the Barmy Army – something to take pride in off the pitch, as there wasn't much to be proud of on it.

This sort of thing really can't be manufactured. It's all about listening for and seizing the opportunities that are created everyday by all the people using social media.

The opportunities also mean that many more business functions should play an active role in listening to social media. You can discover business intelligence and conduct market research. Social media plays a crucial role in customer service, technical support and consumer insight management. It is key to reputation and crisis management. It can find new talent to hire and new facilities to expand into.

This is all part of the humanization of business. We must interact with customers to solve their daily challenges and assist them in reaching their goals. Along the way, there will be many opportunities to exploit, but first you must tune in, turn up and stand ready.

Don't Overestimate Your Intimacy

At a real cocktail party, assuming you have not had too much to drink, you can tell how close a connection you have with someone. Face-to-face, it is easier to gauge when they are hungry for more of you, and when you are coming on too strong. I say easier, because even face-to-face this is something we can all get wrong sometimes.

Online something different happens. The distance may be closed easily, but that should not be mistaken for intimacy.

People will often have relatively close conversations online with people they barely know, and to newcomers to social media this can feel like a powerful personal bond. It's weaker than they think.

While the relationships that you make through social media might not have been possible by other means, and you may feel yourself growing 'closer' to people online, you must remember that digital intimacy is not 'true' intimacy. It's just a trick of the brain.

Go ahead, be friendly to everyone. Provide great conversation. Draw people into relationships. But always maintain professional perspective. Until you actually meet people in real life, perhaps at a real cocktail party, don't assume you have made a new best friend forever.

Take Away Express #3

- ☐ Listen for the details. Monitor more than your brand and products. Find the keywords that create opportunities for you to add value.
- ☐ You can't schedule spontaneous opportunity, but you can be ready for it.
- ☐ As business transforms, it will need authentic Social-eaders ready to interact with people.
- ☐ Don't mistake online cocktail party conversations for intimate relationships.

4. ON WITH THE SHOW

'What you are will show in what you do.'

– Thomas Edison (1847–1931)

This is not a campaign. This is your life. You're not going to be able to just 'do' social media for a few weeks or months. There is no escaping this responsibility. Social media is not a campaign, no more than having a phone is a campaign. This is an every day, always-manned, forever and ever sort of thing. It needs to be sustainable.

Social media is not just a cost of doing business, a 'me too' technology that is too often relegated to the supervision of an intern. Why would you treat such a powerful business tool with such lack of respect? This is an opportunity to shine, to demonstrate thought leadership, to show the world what you are made of. It's time to get on with it.

The battle for attention is on. 'Zuckerberg's Law', first coined in 2008[1] and still a bit too new to really be considered a law, shows that the volume of photos, status updates and other online material each individual posts on Facebook has been doubling every year. That's a lot of social currency.

So far the numbers are already staggering. Facebook users interact with 900 million 'objects' which include pages, groups and events. Facebook users trade 7 billion pieces of content per week, including 250 million photos per day.[2] And that's just Facebook.

At no point in the future will you be able to afford to share less than you do now. Socialeaders need to be personally alive and active in this social economy. They must think like a media company, becoming publishers, editors and the stars of the show.

More and more communities want to hear from a person of authority within a company, not some intern who has been assigned to tweet. When we have problems with a product or service, we all want someone who can truly do something to resolve the issue, to make things right again. We want someone who has the expertise to understand a complex issue. We want someone who can humanize the business.

$E = mc^2$ – Everyone is Now a Media Company

I first read about the idea of all businesses becoming, to some extent, media companies in Clay Shirky's 2008 book *Here Comes Everyone*, where he argued: 'All businesses are media businesses, because whatever else they do, all businesses rely on managing information for two audiences – employees and the world.'[3]

The concept was re-introduced to me a couple years later by Tom Foremski as EC = MC: every company is a media company.[4]

To continue in the tradition of taking this piece of social currency and improving on it before setting it free again, I suggest we evolve this term a step further.

We are all faced with a social media future that is inescapable. There is nowhere left to hide. So we had better decide right now to start personally acting like a media company, to start building not just the media empire for our business, but also for ourselves. That why I put Einstein's famous formula back to rights, although with a new meaning: $E = mc^2$.

The exponent reflects just how much work it is running *both* a business on social media and your own professional profile. Don't underestimate how much time you need to start finding for this.

Start by using this equation to change the way you think about social media in your life and your organization. Use it to elevate the importance of social media in your mind. Social media is not an add-on or an after-thought, they're a key part of the value you add to work and to yourself.

This will call for a change in how we work and in how we organize business. This will call for adopting ideas like 'Brand Journalism'. It makes sense to become our own media companies. In the mass media era, the barriers to entry were too high; it was easier just to buy space in someone else's media. Brands have long been a part of the media landscape. And now so will the individuals who shepherd those brands.

Being a media company is one thing, being a journalist is another. Journalism is not something I treat lightly. I went to journalism school. I worked on magazines as a journalist. I know in my heart of hearts exactly what it means, in its purest form, to be a journalist. I'm an adoring fan of great journalism, and of great journalists. I also know that the real world of journalism often fails to match up to our romantic ideals of the profession.

Journalism, at its best, is all about trust. So is your thought-leadership on social media. The thought-leadership of those who are charged with being custodians of their brands will play just as important a role in the new media mix as journalists did in the old. Brands that hope to be considered 'authentic' have to learn what this trust thing is, and start embracing it.

Brand journalism is not marketing per se, although some may call it content marketing. To get an idea of what good brand journalism should feel like, think what it's like going to a conference and listening to a great speaker talk about a hot topic. Nothing ruins their presentation more than when they start selling from the stage. A great speaker would never do that. Nor would a great journalist. Journalists talk about issues that matter, not about the advertisers that pay their salaries. It's all about sharing value.

When you stop selling from the social media stage and start caring about providing your audience with value, then you are truly beginning to compete against all the noise. You have to prove that you are a more trustworthy source than others. And you gain that trust through the stories you share.

It all comes back to your social currency exchange. In journalism, it doesn't matter if the art department dresses up a story with great images and graphics; if the story is lousy, it fails. You could print it on magical strawberry perfume scented paper, and it's not going to improve the value of the social currency – nor is it going to make me trust the author, the editor or the publisher. Quite the opposite.

The same is said to be true in music: all the explaining in the world won't help a bad song sound good. A good song needs no explanation.

So forget about strawberry flavoured paper gimmicks. The story is everything. When it's authentic it needs no window dressing; when it's false, all the gimmicks in the world won't help.

Socialeaders who can trade in authentic stories of value, who can think like editors and like publishers are going to be in high demand in the years ahead. While this book isn't about selling you on the organizational storytelling idea, as a Socialeader this is going to become an obvious part of the transformation that lies ahead for you. More and more organizations will sustain themselves as a result of the stories we tell each other. We really do all need to think like journalists.

Brand journalists already exist. They come in many forms and likely don't even identify themselves as such. But they are good at telling brand-related stories in a way that can easily be remembered and passed on. They are good at finding the emotional charge that makes a story valuable social currency.

Steve Jobs, Apple's co-founder and CEO until he died in 2011, could do this sort of journalism. He could take all the complexity out of his business and give you the humanized nugget.

When he revealed the first MacBook Air, he pulled it out of a manila envelope. Apple didn't concoct some sort of unusual sized envelope. It was something real, something authentic, and something we can easily understand. It was so thin that it fitted into an envelope. Wow!

We all need to tap into the spirit of Steve Jobs as we set out to become our own media companies.

What Matters at Work?

In order to tell the most valuable stories, you should seek to understand what is important at work. What are the values of the culture, the objectives, the KPIs (Key Performance Indicators), the mission, the vision? These are all things that so many consulting firms are set up to help you with, because for many organizations defining these things is not easy.

Where I've seen this go wrong is when the consultants cook up some amazing values that work as a marketing tool, but don't reflect the actual values of the people who work for the company. What matters at work can't be something dreamt up or concocted to aspire to.

When you are clear as a Socialeader about what the values are, then people can begin to connect with you for the right reasons. And it's not just about connecting with people out there in the wider world; you also need to connect with your fellow employees.

We have long shared social currency with our fellow employees at work. We have always built social capital this way at work. These mechanisms have been a part of every organization for a long time. Now they are digital, and increasingly transparent to the outside world. The stuff that excites employees and gets them talking can sometimes be at odds with the brand's carefully manicured external image.

I once spoke before a group of new managers for a large regional brand in Asia. As my presentation moved into discussing the new era of transparency, several hands went up. They were terrified by the openness to which social media would expose their brand. Their gravest concern was that the world would somehow suddenly discover that things on the inside weren't as good as their marketing campaigns had led people to believe.

I assured them this wouldn't be a problem as their marketing campaigns hadn't really been fooling many of their customers for a long time. Their customers were already talking about how things truly were. A quick Twitter search demonstrated my point. No one in that room had ever searched out the conversations that were already happening around their brand.

The issue here stemmed from a mass marketing history of wanting to cover up the failings rather than drag them into the light and make the company better. Reconciling the way things truly are with the way you've been claiming them to be will no doubt be an unsettling period. But this is the revolution that we cannot escape. This is the culture shift that some are calling the 'humanization of business'. It's about keeping it real.

Start clearly defining what really matters at work. Figure out what it is that the champions in the company really love about what they do. What is it that they would love to do? What frustrates them? What frustrates the customers? Are they the same things?

Don't try to manage what the public is saying, fix the problems that they are talking about instead.

What Matters To You?

In 1997 I read a magazine article by Tom Peters titled 'The Brand Called You'. He said, 'It's time for me – and you – to take a lesson from the big brands, a lesson that's true for anyone who's interested in what it takes to stand out and prosper in the new world of work.'[5] We needed to think of ourselves as brands. For me, as a young man just starting out in the world, this was a fabulous insight for advancing a career. Fast forward to today, the idea of a 'Brand You' still resonates with me more than ever in this social media era. We all have social media accounts that we can 'brand'. We all have every opportunity to package, present and market who it is that we are.

Socialeaders must know who they are. They must know what they are about. They have established expertise and authority. They must build loyal, trusting communities on the back of their own social media empires. Socialeaders have clearly become media companies, even if that's not really what they do to pay the bills.

Being your own brand is a must, even if you work for another company. If 'Brand You' is still a new concept to you, then start by asking yourself some of the same things you were asking about work.

What matters? What do you want to achieve? Write down some goals. One of those goals must be to get better known, professionally, online. You must include, at least as a means to that end, the objective of increasing your digital footprint, and elevating your online influence scores.

But what else matters? Look at what you enjoy reading that is related to your work, to your expertise. Look at the sort of things you get excited about and email them to friends and colleagues. You are what you read. You are what you share.

Newsreaders are the key to building your personal brand. If you don't use one, now is the time to change that. Newsreaders like Google Reader let you explore vastly larger amounts of important news, putting even more social currency at your fingertips.

Certainly, what matters to you will also matter to other people – even if it takes a while to find them. Start by sharing what you read, and only read the stuff that matters most.

Overcoming Stage Fright

Shakespeare understood: 'All the world's a stage, And all the men and women merely players'. It is time to play your part on the social media stage.

But nerves are something hard to overcome.

It's not unusual to feel a rising sense of apprehension as you move towards establishing an active online presence. There is the fear of saying something wrong, the worry of reprisals from imagined enemies, and that sense of foreboding that you are somehow about to be stripped of your privacy.

These worries are all things in your head. And feeding these worries will prevent you from using social media to achieve you goals. You can't be frozen with fear. You can't be stiff. And you can't let yourself prevent yourself from succeeding.

I once heard the comedian Bill Cosby recount his first big break of the early 1960s.[6] While working as a comedian in New York early in his career, Cosby had been scouted by the Marienthal brothers who owned the legendary club, Mr. Kelly's, in Chicago. The brothers offered him a big salary, a plane ticket and a chance to be on one of the most important stages for comedians at the time. He jumped at the opportunity. On his first day at Mr. Kelly's, he went to the club early and marvelled at the pictures of all the famous funny people on the walls. And slowly he began to worry about how he stacked up against the famous people that were once on Mr. Kelly's stage. These were his idols: how was he to compete? His confidence began to sink. By the time he was called on stage, he was at rock bottom.

He took the stage and did his 35 minute act in 18 minutes. He bombed. He went back to his dressing room and thought, 'This is the end.' He apologized to the Marienthal brothers and prepared never to return. He'd blown it. Fortunately, the Marienthal brothers knew show business, and they knew that first performances aren't always the best. They told him that he should definitely go away, and that he should send the real Bill Cosby to do the second show the next night.

When he returned the next night, his confidence was still nowhere to be found. But this time it was different. This time as he took the stage, the owners used the opportunity to poke fun at Cosby, which he responded to by throwing a joke back at them. Then a few more humorous quips were exchanged.

The audience laughed, thinking it was all part of the act, but it had been unscripted. The witty exchange helped loosen Cosby up, and allowed him to feel the energy of the audience. That night he killed them. He did his 35 minute act in an hour and ten minutes. And his career never looked back.

As a Socialeader, you need to send the confident versions of yourself out onto the social media stage. Don't talk yourself into not being you. Don't let nerves undermine your expertise. Don't let your worries keep you from wowing the audience. Expect that the beginnings aren't going to be easy. Know that everyone at some point has their 'ah ha' moment. It will happen. Something will click and you will be dazzled and amazed – and others will be dazzled and amazed by you.

And don't worry about something really embarrassing happening. It will. But it won't be as bad as you might think. There have been a lot of very high-profile celebrity blunders in the past few years (and likely still many more to come). On top of that, there are countless more low-profile embarrassing moments among us less famous folk. The good news is that this is happening to everyone, which makes us all that much more sympathetic when it happens to someone else.

I call it Mutually Assured Humiliation.

We all need to be able to laugh at ourselves. Don't take 'Brand You' too seriously. You will be all the more loved if you know how to make fun of yourself. It shows confidence. Don't be afraid to play your part.

No One Can Do This for You and You Can't Be Automated

One of the things I love about stand-up comedy is the spontaneity, the way a comedian can interact with the live audience to produce results that no one could have scripted. The same is true for social media.

You need five things for a great 'live show' in social media: immediacy, personalization, interpretation, authenticity and the ability to seize opportunities.

But one of the first objections I get from many people, especially those in senior business roles (those with the greatest potential to be Socialeaders), is: 'I'm too busy to do this myself.' They want to try to automate or outsource their 'live' social media efforts.

I usually advise against it. There is some opportunity to schedule some updates and auto-crosspost to multiple social media accounts, but for the bulk of the storytelling, you should be the person doing it.

Yet, busy people still look to installing 'bots' (computer programs with instructions on what sort of content to post) to do the curation. This, without final human approval of stories before they get posted, is reckless. So many embarrassing, off topic or spammy items can slip through.

The other option is hiring a ghost writer, which seems like a clever way to ensure quality stays high, but this can also be a dangerous strategy in a world where people can smell the inauthentic in a single whiff.

Here is what to consider before farming out your social media empire building:

1) Immediacy: There is only so much someone else can say on your behalf before they have to check with you. Spontaneity is killed.
2) Personalization: Ghost writers can ape your writing and expressions for a while, but eventually it becomes more about them than about you. The way you behave online is as individualistic as your fingerprint. Only you can give it your own spin. Maybe the ghost spins better than you. Then one day they quit. The discrepancy will be noticed.
3) Interpretation: As an 'expert' are you really willing to risk your reputation on analysis done by your ghost? Better hope they know as much as you. If they do, they may be setting up their own social media empire for 'brand them'.
4) Authenticity: People like knowing they are getting the real thing. Trust is everything, and not something you should risk.
5) Opportunity: Opportunity needs people to exploit it. When someone makes you an offer, you need to be there. When you rely on a bot, this never happens. Bots are incapable of doing anything meaningful.

That said, each Socialeader's situation is unique. If you can find a balance between being there yourself, with a bit of augmented support that's carefully managed and groomed, then go for it. It's your audience that will be your ultimate judge. I know that I'd rather see the real you.

Take Away Express #4

- ☐ Not a campaign. This is always on – for the rest of your life.
- ☐ Start thinking like a media company. Elevate the importance of social media in everything you do.
- ☐ Know what matters at work:
  - know your Key Performance Indicators, mission, vision, culture, values
  - bring that to your social currency selection.
- ☐ Know yourself:
  - what is success to you? (It can't be to get more followers)
  - you are what you read. You are what you share.
- ☐ Remember Bill Cosby: don't intimidate yourself. Be confident about what you're good at.
- ☐ Say no to bots and ghosts: you have to be present and do this yourself.

5. WHOSE STORY IS IT?

'You have to know exactly what you want out of your career. If you want to be a star, you don't bother with other things.'

– Marilyn Horne[1]

Brands need storytellers, but they must accept that they don't own them. It has been said that the most important resource your company has walks out the door at the end of every day. Sometimes they don't come back. In the social media era that can mean they leave with a lot of the online relationships they have been 'managing'.

But this was the case even before social media. When a client-facing employee departed a business, despite the legal department's best efforts to constrain them with non-compete clauses and the like, that employee always took a certain amount of the business's relationships with them.

This is the way of relationships. People are loyal to other people. Now as a Socialeader, when you share the social currency of your master narrative, it will attract a community not only to your business, but also very likely to you too. It's easier to form relationships with other people than it is with a brand logo.

This poses a difficult dilemma for many employers. How to manage the balance and reduce conflict? Who should own the storytelling accounts?

The answer is: it is both your story and the company's story. It's a bit like yin and yang. Neither is a better path, so the wise must find the middle way. The good news is that social currency is very adaptable. It's not hard for one story to serve two masters.

As a Socialeader, your face and name need to be seen. But as a business person, you need to ensure that your online efforts are also benefiting the company. The success of one is tied to the success of the other. Ultimately, however, the relationships and trust you are going to develop are going to have to be personal.

Using Social Media at Work

Hopefully, I don't need to get into the argument as to why social media shouldn't be blocked at work. If you still work at a place that prohibits the use of social media, then you need to start a revolution.

Let's be clear: before you can do anything to help your business, the business needs to be ready for social media. Social media policies need to be in place. Training needs to be provided to everyone on what is expected. If your business is just starting out, careful monitoring and mentoring of social media practitioners are crucial in the early days. It's not all that different from giving everyone a phone, or giving everyone a computer with email. We've introduced new communications to the workplace before and there are always growing pains.

Studies show that, as a Socialeader, your online behaviour plays a key role in setting the standard for the rest of the workplace.[2] Lead by example. Demonstrate how to act professionally, just as you would with the phone or email.

There is very little leeway in this radically transparent world to conduct yourself in a public fashion that may damage your business. Purely by association, if your bad behaviour online offends someone, they certainly won't want a relationship with the brand you represent.

As Warren Buffett famously said, 'It takes 20 years to build a reputation and five minutes to ruin it.'[3]

While every workplace is different, the common challenge for all will be how to foster appropriate online activity while still respecting employee individuality and personal expression. Managing potential conflict between the individual and the brand requires clearly defined responsibilities that everyone can agree on.

Rather than banning social media, we need to provide tools that allow the individual's brand better to align with their corporate brands. We need to humanize the corporate brands. We need to create value-based workplace culture. We need to understand and share in those values. We need to walk the talk.

If the values at work aren't a part of your values, then you likely need to find some place better suited to you. If that's not possible, then perhaps you're a candidate for maintaining a secret identity.

He Who Controls the Account, Controls the Story

There are four choices when it comes to deciding how to set up social media accounts at work:

1) Faceless Logo: Years of branding experience lead many to believe that this branded account is an important property to claim. Certainly we wouldn't want our brand to fall into the hands of another. That doesn't mean that this needs to be the primary face of your online activities. People simply dislike interacting with logos – it's a bit like doing business with a big fuzzy mascot and not knowing who is inside it.

2) Brand_Named Employees: Early adopters like Zappos.com, the online shoe and apparel company, went down this route giving every employee an @Zappos_name Twitter account. This keeps all control of the accounts with management. Employees don't even need to know the passwords to the accounts. Theoretically all relationships remain with those accounts when an employee leaves.

3) Individuals: Personal accounts without any alignment to the brand. Often these exist when employees are banned from social media at work, or when they are asked to work through either of the first two types of account. This breeds situations where conflicts can arise. While we can't demand that employees stop using social media in their own time, we can provide them with better tools.

4) Brandividuals:[4] These are the accounts of Socialeaders. They place their work persona first, but also allow for plenty of personal interactions. There is no mistaking that the account belongs to an individual. And there is no mistaking who that individual works for. Control of the account is in the hands of the individual.

Bigger companies are providing new tools to help more employees become brandividuals. For example, Hewlett-Packard created what they call a 'centre of excellence' or CoE to serve as a guiding light.[5] The idea behind a CoE is to provide social media users in your company with on-going training, oversight and best practices.

When Hewlett Packard's CoE first began, the company had a very limited awareness of how employees were using social media. What they discovered was that there were lots of branded and brandividual accounts already out there – many that the company had no idea about. From there, HP put a governance model in place, not to slow their employees down, but rather the opposite – to help keep them going. HP Enterprise now has more than 350 employees using social media, and the CoE is helping employees get better at helping customers.

If your company is big enough to support a CoE, it can help employees by taking care of the more technical tasks behind social media (like analytics), so employees can get on with serving their communities by sharing the stories that they know matter.

Help Yourself while Helping the Business

Of course, not all of us are going to work for a multinational brand with a social media CoE. Many employers are still going to demand that social media is managed from behind a faceless logo.

All the more reason that you should never give all your curation time and energy over to the brand's account. You are not going to get it back. You will still have to be you for the rest of your career. You need to ensure that your thought-leadership is protected, but you don't have to work against the brand or the business. You can do both things at once.

The tools are there to manage multiple accounts at once. Look for a social media suite that allows you to see all the activity from both your account and the business account in one place. A few examples are Tweetdeck, Hootsuite and CoTweet, but this is a rapidly evolving space so keep an eye out for new innovations.

Through these tools, it's possible to decide whether the same social currency that goes out on one account should also go out on another. As the tools evolve, we are increasingly able to decide which 'circles' of 'friends' we share which stories with. This is great for those who want to vent some personal frustration about morning traffic with just close colleagues via a personal account, and then send out the top industry headlines of the day to all the brand's fans.

Social media suites make this easy to do. With these tools you can serve social currency many different ways to many different people from many different accounts. But keep in mind that this also opens up the risk that at some point you make a mistake, just as with managing a secret identity using the same tools, and accidentally send something out to the wrong group via the wrong account.

The best strategy is still to be working to align all the accounts around common values that you are truly passionate about.

Only the Passionate Enthral

Passion can't be faked. There will no doubt be various levels of passion within any organization. Those with it are your company's natural storytellers, even if they don't work in the marketing department. Even if they have never blogged or tweeted before.

The modern business storyteller needs to be someone who has the instincts of an editor. It's about judgement: recognizing when something is hot or not. They are also usually 'true believers' in the brand values (even if the company doesn't always live up to them). They are passionate about what the company does right (and often critical of what it does wrong).

But 'creating content' often meets with resistance outside the marketing department. Those who've tried to roll out 'content marketing' strategies have found most non-marketing people will offer excuses like:[6]

'I don't have anything to write about.'

'I don't have the time to write.'

'Isn't that the marketing department's job?'

Knowing that customers prefer to talk to people in the company with authority,[7] it's not critical that you get every single person in the business using social media. Focus on those with authority and passion. These are the people to get building up themselves and the business online.

If someone has passion, provide them with all the other support. That's where strategies like a CoE come in. Be that resource for others. Help show them how easy it is to do. Show them how the tools work. Be there to answer questions as they get started. It is much easier to do all that than it is to teach someone to be passionate about their work.

Reputational Risk: The Door Swings Both Ways

For me, there is nothing worse than going to a conference and then having to listen to a speaker sell from the stage. I refuse to believe a word they are telling me. The truth is that social media are just like the stage at a conference.

So, when you are sharing your social currency, ask yourself: are you being paid to share that link, or are you sharing it out of passion? Can you do both? What does the word integrity mean to you?

If you think you can sneak one past us, think again. The consumer's BS detector is better than you know. But there is an acceptable way to share links to stuff that you are getting paid to promote and still retain your integrity.

Disclosure is the first step. Tell us up front that you are receiving some sort of remuneration from those behind whatever it is you are talking about. Everyone will appreciate that you are being honest from the outset.

It's then up to your passion to win us over, to make us feel that you genuinely endorse that which you are sharing with us for reasons other than the money.

Keep in mind that as a Socialeader, everything you share and talk about is also affecting your reputation. So while brands worry about employees tainting their reputation, you need to be worried about the brand tainting your reputation.

Businesses sometimes get embroiled in controversy. This means that you as an employee get dragged into the muck too. If the business pushes you to tow a line that you know to be untrue, and the truth comes out (and it always does), then your personal brand will suffer too.

Your reputation is something you need for the rest of your life. Be careful how you let those you work for put it at risk.

Take Away Express #5

- ☐ Conflict may arise between the company and individuals over the proper use of social media. The solutions require:
  - clear policies, positive role models and centres of excellence
  - culture shift, humanization, living the values.
- ☐ Controlling the social media accounts: claim the brand, but empower brandividuals to manage their own accounts.
- ☐ Help yourself. Use social media suites to put your thought-leadership in multiple places at once.
- ☐ Passion can't be faked. Find those who truly care and teach them the social media tools. It's easier than finding those who know the tools and trying to teach them how to truly care.
- ☐ Don't sell out. Integrity matters. You are stuck with your reputation for the rest of your career. Don't let a brand drag you down.

6. HOW TO NEVER RUN OUT OF INTERESTING THINGS TO TALK ABOUT

'Why create anything new when there's a mountain of freshly excavated pop culture to recut, repurpose, and manipulate on your iMovie?'

– Patton Oswalt[1]

There is a glut of 'content', but valuable 'currency' is always scarce. No one could possibly read everything that is being produced. Google's former CEO Eric Schmidt claims that every 48 hours we produce as much information as was produced from the dawn of civilization until the year 2003.[2] And we can expect that trend to continue to accelerate.

But despite all this information, social currency is still scarce, and you're going to need to be trading a lot of it every day. You can create some of it yourself, but the rest of it you will have to go out and find.

The early days of the social web have taught us that it's easy to start out strong, writing blog posts every day. But, more often than not, that pace slows to a crawl. It was once said that everyone had one novel in them. Now I think we all have one great blog post in us. After that, it gets hard. The web is littered with the untended remains of thousands and thousands of blogs and social media accounts that began well then ended suddenly.

The challenge for most beginners is finding 'original' things to write about. But that was their big mistake. Blogs are by their very earliest definition logs of the web – collections of found things that the blogger thought worth noting. So, what we really need to be doing is looking for 'new' things to read about. Where to look? Across your industry, sector, consumer segments. Look into related fields, look to academics, look to the lunatic fringe. Anywhere that you can find relevant news, stories or research that your community will enjoy and want to share. Getting good at finding new stories is what will make you a great curator – and keep you from running out of things to talk about.

Curation is much easier than creation. You should still try to create original currency whenever you can, but in between you should be looking at all the other stuff you can share via social media channels.

Curation shouldn't be looked down on. Its great beauty is that it doesn't add more noise, but instead works as a filter. Curation is you being the editor of the narrative, using your expertise to choose hot or not for the rest of us.

To be a great curator you must first start by becoming a master of your newsreader. The top Socialeaders today have been using newsreaders since RSS arrived in the blogosphere around the turn of the millennium. For many it seems to be one of the best kept secrets of the modern workplace.

Instead of wasting your valuable time going to check a few dozen news sites, newsreaders allow you to subscribe to hundreds of sources. This helps keep your curation fresh. You don't want to always collect the same stories you see others in your community collecting. The social currency will be more valuable if it is fresh.

Work Smart: Combine Your Read and Write Time

You need to read the news anyway. Why not use tools that make both reading and sharing super easy? Why not use tools that are already connected to your social media accounts? Stop consuming news any other way and become a newsreader ninja – you'll consume far more, far better, in far less time.

All the best newsreaders are now fully integrated with your social media channels, so it's a matter of a click or a tap of the screen to share a story that you think is valuable social currency. Many of the new newsreader apps use Google Reader as the backend platform, so I'd recommend setting up a Reader account if you don't have one.

Then look to tools that make Google Reader even better, like Feedly for your browser, tablets and smartphones. It makes all your blog subscriptions look like a magazine layout, making it really fast to scan the news and select what's hot or not. With another click you can share it with your community. Fast and effective – that's smart.

Never again waste time going from site to site looking for news. Apps like Flipboard, Zite and Pulse can turn an iPad into the ultimate news curating machine. They allow you easily to subscribe to, manage, organize and filter hundreds and hundreds of sources of social currency – all laid out beautifully like your morning newspaper. While you are waiting for your morning coffee, or while you are commuting to work, you can easily be sharing the top stories of the day.

You can be curating while those with newspapers are still just consuming.

Newsreaders, at first, need you to tell them what content to go out and collect. Where to start subscribing? Most top newsreaders now have a built in recommendation engine. All you have to do is start typing the topics that you are interested in, and it will show you related blogs. Often the newsreader will also show you how many subscribers each blog has, which helps you decide which are best to start subscribing to. Start with the biggest; they add value by layering on opinion and insider coolness to the news of the day. It's easy to add dozens and dozens of new sources in minutes this way.

Most social media build upon traditional news sources,[3] and so should you. Traditional news sites make an easy first add to your newsreader: you were probably visiting these sites anyway, now they will come to you. Many top news sites offer RSS feed for each of the various departments (i.e. world news, politics, science/tech, lifestyle). Subscribe to just the departments that matter to you and your work.

Then there are the other blogs in your industry. Look for your competitors. Add them to your newsreader too. Know what everyone is saying.

Then get social. Add person feeds from top writers and editors at the blogs and news sites – these are the people you want to start building relationships with.

Then start adding less well known sources. Look for some fringe stuff. Look for some funny stuff. Look for dissenting views. Add lots of variety to the mix.

With many of these new reader apps, the more you read, like and share, the smarter they get – assisting in surfacing even more valuable social currency. These apps are also social graph enabled, which means they can learn what your important network of 'friends' is reading and liking. This boosts the quality of recommendations the apps make.

This is a rapidly evolving space, so try a few until you find one that you really like. Always be on the lookout for any tool that makes the curation process easier. This is the sort of advantage you need as a Socialeader.

Once you have the tools in place, you can then start to practise the different types of curation.

Distilling the Master Narrative

One of our greatest challenges is to overcome our mass media mentality. Stop trying to sell yourself. Start sharing more about others. Aim for a ratio of about 1:20 – one story about your brand for every 20 other interesting stories you can share. To achieve that, you'll need your newsreader to be tuned into your master narrative.

The master narrative in which I reside is the world of geek news. It's a narrative where tech announcements erupt in my newsreader daily like volcanic plot twists in a J.J. Abrams thriller. I follow and share in the battles of Apple, Google, Amazon, Microsoft, Facebook, Twitter and many, many more. It's a noisy place where it always feels like I'm surrounded by people who know so much more than I do. Yet, outside the tech set, there is this blissful ignorance about this narrative and the drama being experienced by those of us on the inside.

Outside of my master narrative, people barely know what a 'meme' is, let alone what things like Sad Keanu, LOLcats or Domo Kuns are. The social media realm is often described as an echo chamber, where issues and topics of the day seem much, much louder on the inside.

The deeper you can get into the 'geekiness' of your master narrative, the more value you can bring back to your broader community. It's about mastering the material: knowing the inside scoop, listening to the minority who are even 'geekier' than you discuss the minutiae of the latest and greatest thing, then presenting those ideas in a manner that is easier for the less geek majority in your community to comprehend.

This type of curation is like distillation: taking the complex and refining it into something easier to appreciate. This is a great way to create more valuable social currency without having to actually create any content.

It shouldn't take an ubergeek to realize that distilling your master narrative makes your tweets and Facebook updates so much more valued by online communities than if you post traditional marketing messages.

Build on Other Awesome Curators

While it is relatively easy to perform distillation curation on the fly as you are reading through your newsreader throughout the day, there are other forms of curation that will add even more value to you social currency.

The deeper you read into your master narrative, the more your expert eye will identify trends. You will start seeing similar stories emerge form very different places, at about the same time. It's at those moments you know something bigger is happening. Don't underestimate the value of being able to connect the dots for other people.

Gather together the various small sources and turn them into a big source. Use your blog to explain the relevance. This isn't really 'original' content. You are using mostly social currency from other places, but this will be much more 'sharable' than if you just shared links to each story without connecting the dots.

There is another type of mash-up curation which does the exact opposite. Instead of putting similar ideas together, juxtapose opposites. Make sure your newsreader pulls from a wide enough variety of sources that you see thoughts and opinions that are unpopular – even if you don't agree with them. Then when you see other curators sharing conflicting social currency, pit these ideas against each other in a battle royal on your own blog and let your community of commenters determine the winner.

While both of these techniques require a bit of re-packaging work, it is essentially still just curation, nothing needs to be made from scratch. Yet, the effect will be that those who see what you've done will see that you have an expertise in the area.

When all else fails, round up a Top 10 list (or even just a Top 5). These make for great social currency. People love lists.

But whatever you do, don't steal from creators or other curators. Curation is not about theft. It's about helping drive attention to places that deserve it. If you share more than a headline and a link, then make sure you tip your hat to all of your sources. Remember to thank people, link to them, mention their names. Make sure that they know that you are a fan. This builds a relationship and makes you a valuable part of the community.

Likewise, gratitude also needs to go in the other direction. When they share something you've put together, thank them for spreading your currency. This is how we build our social capital: mutual exchanges, and acts of gratitude.

This is stuff that cannot be bought. This is truly earned.

Groom the Stable

Always be on the lookout for new sources to add to your newsreader. Look who other influential social media users in your master narrative source from. Go directly to those sources and add them to your stable.

Don't discount blogs on the fringe. They too can provide unusual uniqueness, which can be valuable. They can also rise.

New more authoritative voices are always in demand. Nobody wants to have to share from the same guys all the time. Variety is the spice of life. Changing things up, looking for new talent, revisiting old friends. This is an important part of curating.

So is cleaning house. It's quite easy after a few months to find yourself subscribing to way more stuff than you are reading. You may find yourself reading the same few blogs that are at the top of your reader and running out of time to get to any others. Reorganize your feeds to keep a fresh source of social currency flowing through your day.

The only way to keep up with the demands of a Socialeader is to keep your own interest levels up. Find funny stuff. Share the lighter side. Don't let yourself be one of those people who only had one good blog post in them.

Don't Get too Hung up on 'Signal', Make Some 'Noise' too

I remember in Twitter's early days being confronted by business people who would dismiss the social media service as nothing more than 'people talking about what they had for breakfast'. I enjoyed reminding them that it's not about knowing what everybody had for breakfast, it's about knowing what your VIP client had. The same people who dismissed Twitter could hardly dismiss the value of knowing the needs, likes and desires of their key customers.

The point is that in business small talk does matter. When you have a meeting with a key client, you always begin with some pleasantries and small talk. It's a social currency exchange that helps deepen the bond and lubricate the negotiations ahead.

Saying hello or thank you certainly doesn't qualify as content, but it is social currency. Pleasantries are emotionally charged. A 'Hello' reciprocates a 'Hello'. A 'thank you' often elicits a

'You're welcome'. The same happens in social media. This is an important part of the humanization of business.

Having conversations with your social media community isn't really 'content creation' nor is it 'curation'. Neither is providing friendly customer service. These things are classified as non-content 'noise' and, despite what many will say about it, it's often more important than a lot of 'signal'.

If you get hung up on providing only the 'signal' then the danger is that you appear 'anti-social'. Don't be deaf to how the 'noise' is actually building relationships. Relationships are the whole reason we are using social media instead of billboards.

Noise can be good. Not every social currency exchange requires a link, a photo or video. Sometimes it needs to be about injecting personality. The occasional personal comment or joke that relates to stories that are well known within the master narrative will be well received.

People bond more through small talk than we think.[4] Watch for appropriate opportunities to make small talk. That's the sort of 'engagement' that really builds trust, which is one of the keys to becoming an influential Socialeader.

Take Away Express #6

- ☐ Your master narrative: it's a river of social currency for you to filter and share with your community.
- ☐ Master the newsreader: work smart, start with Google Reader, search YouTube for lessons on how to use, read and share from the same window.

- ☐ Get inside the echo chamber: learn from those who are experts, help those who are learners.
- ☐ Add value by re-packaging social currency: put like ideas together, pit opposing ideas against each other.
- ☐ Keep it fresh: look for unusual stories/perspectives that your community hasn't seen yet.
- ☐ Don't steal: curation must always give credit where credit is due.
- ☐ Not just the news: small talk, pleasantries, asides and other chit chat build social bonds too.

7. DON'T JUST TELL. PERFORM!

'If I have seen further than others, it is by standing on the shoulders of giants.'

– Sir Isaac Newton

We must sing for our supper, so to speak. Audiences are drawn to, and are most likely to share, the sort of story that is imbued with human character. We treasure a storyteller with flair, style, wit and charm.

None of that is easy to learn. And it's impossible to automate. Remember, we need to offer immediacy, personalization, interpretation, authenticity, and have the ability to seize opportunities.

Instead, to prevent the demands of social media from grinding us down, it's important to maintain a balance. It's not unusual for people to set aside certain 'office hours' when they are 'live' on social media. Also you need to make it clear when you will be unavailable.

Ultimately, it is a marathon not a sprint. Social media is about sustainability. You should try to shape your social media activities into something you enjoy doing over the long term. Communities will connect with our genuine passion, so spend as much time as you can on topics you really care about.

As you begin to build a following, it gets easier. The feedback from the audience makes it all worthwhile. When you start seeing other people pass along your social currency it encourages you to want to share even more. The first time you have conversations online with a person you admire and respect, whom you never would have had a chance to speak to in real life, is a thrilling experience.

Feed off the energy of the audience. Sense the crowd, and adjust your performance like a DJ changing records to keep the energy up all night long. It will also keep your enthusiasm up as you begin your lifelong journey as a Socialeader.

Live from New York, it's Saturday Night

As a teenager, I would stay up every Saturday to watch the legendary sketch comedy of *Saturday Night Live*. As would many of my friends. Being pre-Internet, we did what many teenagers would do, we would call each other on the phone. We would laugh and share in the jokes together. We shared this 'live' appointment together, and it was more enjoyable because we shared it.

Being present in a moment is great, and there are more opportunities than ever to be there together, virtually, with friends when something big is happening. Live streams pour out from conferences and from big technology product announcements. I'm often up in the wee hours of the morning so that I can share in the excitement on the other side of the planet. The bonds always deepen between those of us who also stay up on Twitter together – there's a palpable sense of camaraderie.

Look for these sorts of opportunities to connect live. It doesn't always have to be some major event, or in the middle of the night. It can happen anytime around breaking news or major media events. From royal weddings to massive hurricanes, from fashion weeks to a World Cup sporting match – any major experience that can be shared, even via social media, brings us closer together.

However, the truth about *Saturday Night Live* is that it was, at least for me and my friends living in Western Canada, never truly 'live'. It had been recorded in New York some hours earlier and replayed at midnight in my time zone on the other side of the continent.

Television was all about scheduling. And now so is social media. Your social media suite, like Tweetdeck or Hootsuite, will allow you to schedule tweets to go out at exact times. When you send a tweet can make a significant difference to how much response you get to it. Studies have shown that there is even a 'prime time' for getting the most retweets (around 4pm EST).[1]

But there will be different times for when your community is most active. Services like Crowdbooster.com can analyze your network's responses and provide recommendations on when your social currency gets the most traction.

Many Socialeaders have also discovered what TV knew all those years ago: you need to repeat your show to accommodate different time zones. And as social media gives you a global reach, you certainly want to take into consideration re-running your social currency at a time when the other side of the planet is awake.

The drawback to this is that you will always have some night-owls in your community who see the re-runs showing up twice in their social media streams, and feel somewhat irritated. So you need to strike a balance. You will most certainly get more people viewing your social currency every time it gets repeated,[2] but if you repeat it too much you can lose audience too.

The other problem with scheduling your social currency to be auto-shared while you are sleeping is that you can't really be there to connect with fans as they react. The best you can do is to try to get back to them as soon as you wake up.

One Take Makes Great Currency

One of the first truly viral YouTube videos I shared was OK Go's 'A Million Ways'.[3] The band shot it in their back garden. It was choreographed by one of the band member's sisters. It was just four guys singing and dancing their way through a very complex, and humorous, dance sequence. In one take.

It became a social currency phenomenon.[4] It was copied hundreds and hundreds of times, with each video being put up on YouTube in response. Each video done in one take.

And that was the magic. Not that it was great dancing. It was a bit funny. But the effort that you could see the band putting into it gave it an emotional charge.

They've followed up the 'A Million Ways' video with many more that have been done in a single take. 'Here It Goes Again'[5] had the four band members dancing on a series of running machines. Just one take.

For 'This Too Shall Pass'[6] the band built, with the help of an army of engineering geniuses from Syyn Labs, what may be one of the world's largest and most complex Rube Goldberg Machines.[7] A RGM is a deliberately over-engineered contraption that performs a very simple task in a very complex fashion, usually including a complex chain reaction. This one took a month to build. It was set off by a toy truck colliding with dominoes, which, after dozens and dozens of complicated transitions including rolling car tyres, smashing televisions, runaway cars, cascades of ping pong balls, human slingshots and a falling piano, finally led to the band getting blasted by enormous paint cannons. Just one take – and amazing.

Their 'White Knuckles'[8] video even included dogs. Lots of dogs. Every performer will tell you, 'Never work with children or animals.' It's so much harder. Yet, they nailed it. A massively complex choreographed number. Just one take. Four years in the making.[9]

Every time the effect is the same: the video goes viral.

'One take' makes great currency. But what OK Go teach us, especially with 'A Million Ways', is that it doesn't take a big production budget or special effects – just a willingness to put in the hard work, to practise and practise, until you are capable of truly delighting people.

I've seen OK Go perform live, and their amazing performance demands an enormous amount of effort from the band members, above and beyond the call of duty of most rock stars.

That's how you win fans.

Feedback: Crowd Accelerated Learning

The darkest days are those when you first set out in the social media realm. Everyone starts from zero and it's lonely. But it gets easier – largely because of your fans.

When mankind first began sharing stories, we sat around the campfire and could look into the eyes of the audience. You could tell if you were connecting with them, or if they were drifting off to sleep. The storyteller could speed up, slow down, elaborate more or simply cut to the chase.

It's harder to look into the eyes of everyone who follows you online, but it's a lot easier to get their feedback than in the days of print. Working at magazines I would occasionally get a letter, but it was usually in regard to a story that I'd filed two months before. The delayed reaction, combined with the fact that I was on to other stories, meant that audience feedback wasn't something I could use while in the process of telling the story.

But as a Socialeader, you work in real time. You get better as a storyteller as you respond to the reaction of the audience. Let the community help guide you and build you up. As you share stories, ask for feedback. Take note of where they set the bar, see if you can raise it.

The other trait that has re-emerged from times long ago is crowd accelerated learning, although I'm sure they called it something else back then. Campfire storytellers would hear others retell their stories in new ways, and learn from that how to improve on their own retelling. The same happens in social media, but at an even more rapid pace within master narratives.

TED curator Chris Anderson recently described how an entire generation of talented young people are becoming world class performers, purely by being 'web taught'.[10] They may live on opposite sides of the planet, but they are forming communities of innovators, commenters, cheerleaders and sceptics. They challenge and encourage each other to do better. They recognize that innovation is hard and requires practice. With that comes social status – or social capital – the recognition of the crowd.

What's happening? When young dancers see a video of kids on the other side of the planet performing some cool moves, those who were the audience suddenly become the performers. They practise the new move, innovate and improve on it, then shoot a new video and put it up. Others pick up on that and do the same. And around and around it goes, the audience providing feedback, encouragement and challenges.

Anderson says he's seen the same effect on his TED Talks since beginning to put presentations online – the calibre of presentation has gone up. If you are lucky enough to be asked to present at TED, you need to pack months' worth of preparation into your 18 minutes on stage. Social media performance, likewise, needs to be up to the standards of your community.

So as you first embark on you social media adventures and see a performance amongst your community that wowed you, try to play it back, but even better. Share advice, ask for insights and accept challenges. Take the energy of the audience and put that back into your own social currency mix.

Leverage that to create an even better performance. Use it to become part of the community.

Take Away Express #7

- ☐ Live presence: you need to be present to seize opportunities.
- ☐ Know when to schedule: re-runs can help attract fans in other time zones, but be careful not to irritate fans at home.
- ☐ Hard work is powerful social currency. Give the audience something to marvel at. Challenge them to try it themselves.
- ☐ The feedback will sustain you, and improve you. Get in touch with the audience; they will save you from burn out and boredom.

Part Two

COMMUNITY: WOWING YOUR AUDIENCE

8. FINDING YOUR COMMUNITY

'We are judged by the company we keep. When combined, actions and relationships create a foundation for social capital.'

– Brian Solis[1]

Once you've got your storytelling performance up and running, you won't want to be standing in an empty field. We can't exchange social currency in a vacuum. We need other people. When asked why they use social media, the most common response among active users was people related: friends, connecting, camaraderie, companionship, networking, relationships, audience, outreach, community, water cooler, engagement and rapport.[2]

Despite everyone seeming to be there to make friends, we can't expect anyone to come to us at first – they don't even know we exist. This isn't always a bad thing to begin with. As we begin in obscurity, we can find our footing without anyone really watching. Take solace in knowing your first awkward moments will be unwitnessed. But for Socialeaders, we cannot remain in the darkness. Obscurity is the enemy. Being well known among your community is everything.

In order to get noticed by our community, we need to go seek out the places where they are already hanging out. While finding the community is easier today than it has ever been, there are no pre-made maps to the location of yours. Social media is dynamic, always shifting with the seasons. It's a bit like going on a safari to begin with. You need to set aside some time to conduct the hunt.

You start by looking for the 'Big Five'. Big game hunters in Africa coined the term 'Big Five' to describe the five hardest animals to hunt on foot: lion, leopard, elephant, rhino and buffalo. For your community hunt these are the most important five to start with, not that they are particularly hard to track down: bloggers, professional groups, topic chats, aggregators, and what I call midfluencers.

Before embarking on your community building safari, ensure that you are already posting a stream of content or curated social currency because when you start to get the Big Five to notice you, you want them to be able to check out your stories and see that you have value to offer.

When you do make contact with the Big Five, you will also need to use non-content social currency to begin relationships. Pleasantries and introductions work much better than sales pitches or other demands on our non-existent social capital. Remember, we must deposit before we can withdraw.

On principle, if someone I've just met online asks me to retweet something, the answer is always no (unless it's in aid of a really good cause that I believe in). I feel that it's so rude that they are asking something of me without first sharing something of value, something human, something emotionally charged with me. Actually, instead of asking an influencer to retweet something, just show them something amazing, with no strings attached. There is a much better chance that they will pass it along. Only ask for favours once you've established a relationship. I'm always willing to help friends, but I'm not willing to help those who are only trying to help themselves.

The hunt for community is hardest at first, but gets easier once you've established a presence. Overcoming complete obscurity, moving from being a complete unknown to being a recognizable presence is all it takes to begin to attract an audience. But there is never a time to get complacent. Friends drift apart. You need to be seeking new connections, new opportunities and new voices all the time. If you don't, before long you will again drift back into obscurity and find yourself standing alone in the field.

Where to Start Looking

Community can be found in many places. Social media is continuously spawning new ways for people to connect and interact. The following list is about getting you started. As you get to know the community better, the people you meet will lead you to the people you *really* need to meet. Here are the Big Five:

1) Bloggers

The term blogger once only referred to a fringe group of amateur writers. Now it also includes a lot of professional journalists. In fact, when I talk about bloggers, I'm really thinking about the pros first. But I also need to emphasize the importance of the passionate amateurs – the word amateur meaning to do something for love. The support you get from these 'lovers' will make it easier to keep putting effort into your social media.

Bloggers are the new newsmakers, the new gatekeepers, the new Fifth Estate – the online opposition to the mainstream media establishment. We rely on bloggers for most of our social media currency. These are people you want to build relationships with, gradually. Remember to give before you think about taking. Reference them when you share their stories on Facebook. Cite their Twitter handles when you tweet them. Don't just reference the blogs they work for, especially if they blog for a company or media group of some sort. We want relationships with people, not brands. Drop them tips and insights that have nothing to do with your company. Don't try to sell to them. The more authentic value you share with them the more they will like you.

To find bloggers, start with the obvious watering holes. For years bloggers have checked their blogs into places like Technorati, which makes them easy to search out and see which ones are most influential. You can also just search within most newsreaders, including Google Reader, which will pull from Google's blog search.

There are also sites designed to make finding bloggers easier. BlogDash.com, ecarin.com and grouphigh.com, among others, provide a mix of free and paid services to help find your community. This is not an endorsement of these sites. Your mileage may vary. But they could save you some time.

Bloggers also like to interconnect with other bloggers. Blog rolls, where bloggers list each other in the sidebars of their blogs, have fallen out of fashion, but you can still sometimes find them among smaller bloggers.

Big blogs are often being talked about a lot by smaller ones, and sometimes vice versa. Keep track of who is citing whom as a source: follow those links. Think like a tracker in the jungle, watch for footprints.

While you are looking at blogs, look who is commenting on the blogs. They often have sites of their own. If their comments are interesting, their blogs might be worth checking out.

2) Professional Groups

Within nearly all the big social media sites, you will find professional groups of some sort. I use the term 'professional' loosely. These are groups of experts who often derive some sort of livelihood from that expertise. Groups are for those who are most interested in a particular topic. Most social media platforms, like Facebook and LinkedIn, allow you to search their groups by keyword. Do a search for all the key terms you think are hot in your master narrative.

Not all of the groups you will find are run by true professionals; some amazing groups are maintained by amateurs. Some groups are even run by brands. You'll need to decide which are worth joining. Have a look at how active the groups are. Many get set up and live very brief lives. This goes back to what Dunbar said about gossip, that it provides a maintenance function for groups.[3] Without an ongoing conversation with groups, they cease to function.

Groups that are open to the public often have lots of members, but very few are actually talking or sharing anything. Groups that are actively sharing tend to have more than a few ring leaders. The leaders won't be hard to identify as they will be the ones sharing the bulk of the social currency. Start by befriending these people.

Some groups require permission to get into, which sometimes means there is less noise and the social currency exchanges are often of higher quality. Request permission to join. This is another reason it's important already to have been creating an online presence through the sharing of valuable social currency. When you ask to get into a private group, they will check you out. If you have a record of sharing valuable social currency, it will improve your chance of getting in.

We are judged by our associations. On many social media platforms, like Facebook and LinkedIn, the groups you belong to are listed on your profile. So it doesn't hurt to belong to several of these groups. But don't overdo it. Stick with groups where you find value, and where you can contribute. Be authentic in your memberships.

3) Topic Chats

Topic chats are live, public online conversations. Unlike forums or bulletin boards where people post and maybe get replies days or week (or even years) later, topic chats are real time. These real-time group chats were born on Twitter around hashtags (keywords that are marked with an # symbol), and are now spreading to other social media platforms. These chats provide a live, virtual cocktail party feeling.

There are topic chats on Twitter for nearly every industry these days. They each have weekly time slots when people come together and discuss the topics of the week. A schedule can be found by Googling 'twitter chat schedule'.

The people who host and/or participate every week are again the first ones you should follow and begin building relationships with. But always stay open to opportunity – you never know who might drop in and surprise everyone.

If you do join a live topic chat, remember your non-content social currency: manners matter. There is always basic etiquette: show up on time, formally introduce yourself, always include the hashtag when replying to others in the chat. And have fun.

4) Aggregators

These are giant beasts of curation. They feed off others, pulling together relevant blog content under topic headings. Aggregator sites essentially pull all the RSS feeds from all the top blogs on a subject. You are in good company if you can get your blog noticed by an aggregator, like AllTop or Digg.

The benefit of finding aggregators is that they can quickly help you determine what is popular in topics that interest you. As they have already tracked down a lot of great sources, aggregators provide you with short cuts not just to the best blogs, but to the people behind the blogs.

If you are still trying to build up your newsreader subscriptions, aggregators sometimes offer OPML files (a big bundle of RSS feeds) that you can import to your newsreader to instantly follow all the best sources.

Another type of aggregator is the list builder. Many great lists of influencers have already been put together on Twitter. Have a look at sites like Listorious, which organize the best curated and most followed lists on Twitter. You can simply start following these lists right now, and use your social media suite to set up a column to monitor their action.

5) Midfluencers

Some people in social media don't need a blog to be able to move people. Some of the most connected people have only Twitter accounts. The midfluencers aren't famous people, but they hold a disproportionate amount of online influence. They've usually been using social media for years and along the way built up a trusted network.

When we look on leaderboards like WeFollow.com, it's important not to always focus on those at the very top. 'Most-followed' influencers are difficult to break through to. Instead look way down the list, to those in the middle. They have a few thousand in their networks, but they are clearly passionate about the topics that they are having daily conversations about.

Use tools like Peoplebrowsr or Twitalyzer to explore the networks of midfluencers (or just look through their public streams to see who they talk to the most). Midfluencers tend to hang out among power clusters. These are all people to say hello to and begin to draw into your community.

What to Do When You Find Community

As you find community, the first thing to do is to start following them. I use the word following in the broadest sense. You may need to subscribe or somehow 'befriend' or connect depending on how the social media site works, but the goal is to get them on a list of similar people that you can track using your social media suite.

These lists let you constantly monitor the cocktail party conversations. From these lists will flow the major narrative arc of your master narrative. This is where opportunities will surface.

Add influencers to lists based on who they are and their importance. While you'll want to start finding as many people to list as possible, you'll also need to realize that there is a very real limit to how many of these people you can have a relationship with. The lists will help you focus your community building efforts where they can have the most effect.

Not every social media platform will let you use a social media suite to monitor it. This means you will have routinely to login and keep an eye on the lists and groups that you have formed. Keeping tabs continuously open on a second monitor is a useful strategy so you can check the pulse for a couple of minutes a few times every day.

If you can't watch what the community is saying and doing, then it's not much of a community. So if you want to be part of a big community, you will need to use tools like lists and social media suites that make this possible.

Use Non-Content Social Currency to Build Bridges

Before any of the Big Five will check you out, you need to pay them attention. Attention in today's noisy online world is a high compliment. When you could be spending your time in so many other places, that you choose to focus your attention on someone is definitely meaningful. It's the sort of action that often garners a reaction in kind – they'll pay some attention to you.

Hopefully, they'll like what they see: a Socialeader sharing emotionally charged social currency for the benefit of others. It's not a secret, but it seems overlooked by many community managers: people like to hang out with people they like.

Remember that nothing in the social media era is really done by a brand; it's done by a person working under the brand's umbrella. It's the person you want to find. Think of it as a collection – or a curation – of cool people.

And be cool. Don't be a sycophant. Don't come on too strong. Don't be a screaming fanatic. Don't be a teenage boy and try to close too quickly.

Think about ways you can be mutually beneficial. Don't scheme at ways to make someone do what you want. If you do, you could get yourself blocked. That's the equivalent of being sent into exile. Few ever return from exile. So play it cool.

Bring New People into the Fold

As you begin to find your place within a community, and as you gain a greater share of the narrative, you will begin to attract new people. This fresh blood is critical. Friends sometimes drift apart. There is continual entropy in our social media relationships.

Relationships wax and wane. People move in orbits, disappearing from our radar for a while and then coming back weeks or even months later. The focus of our attention shifts with the current events of the day. And sometimes we just go through periods when we can't put in as much time or effort, so our presence diminishes.

When we meet new and interesting people, we must remember to use social currency that provides them with a sense of belonging. Make it easy for them to relate to the group, to share with their friends.

As your social currency gets picked up and passed along, watch for those who pass it along in the second or third degree of separation. Send them some appreciation, invite them to join groups or follow lists.

Use every opportunity you can to narrow the gaps between you and others in the community. Draw outsiders towards the centre.

Own Your Database

One of the oldest and best pieces of advice from online marketers: maintain a great database of all your contacts. Keep it up to date. Have permission to contact them. Know who matters and how best to reach them. Having each other's email addresses and phone numbers really strengthens community ties.

It's important to remember than many of the lists and groups you create in social media exist within walled gardens – sites from where it is very hard to extract data. Anything can happen to the data within these sites; it is beyond your control.

You want to ensure that you can stay 'friends' with people even if social media sites fall by the wayside. Seek permission from everyone you feel you are developing day-to-day interactions with to collect their contact details. Grow your database as your network grows – it can become much harder to do this retroactively.

Google contact is a good option for storing this information as you can access it from anywhere you can get online and, so far, Google has been a champion of data portability. Your personally owned database will be your central repository of social capital. Don't leave it all behind in a walled garden.

Take Away Express #8

- ☐ The Big Five are gateways to building your community: bloggers, professional social media groups, hashtag chats, aggregators and midfluencers.
- ☐ As you find them, add them to lists that you can readily monitor in real time.
- ☐ Use non-content social currency to foster relationships. Build bridges between others in your community.
- ☐ Constantly be on the lookout for new recruits:
  - sense of belonging
  - relating
  - joining a great cocktail party.
- ☐ Own the database: collect as you go, seek permission, give your details in return.

9. MAKE YOUR 'STORY' FEEL MORE LIKE SERVICE

'The self is not something ready-made, but something in continuous formation through choice of action.'

– John Dewey[1]

As social media is more 'social' than 'media' we must accept that 'relationships' are going to matter more than 'message'. 'Persuasive' marketing has given way to 'caring' marketing.

Even if our jobs are more about marketing than customer service, or vice versa, the future is all about looking for customer problems and solving them.

I grew up in the Big Brand era. And while times are changing, the old way of doing things is still far from dead. I still see it all around me. As do I its culture where employees stay out of reach (and out of touch) of customers. It's a culture that often tries to avoid the customer as much as possible.

One morning I headed down to my local Starbucks and arrived before they were officially open. Staff members were inside, busily preparing the place for business, and the door was unlocked. I went in and asked if there was any chance I could get a coffee. I knew this was asking a bit of a favour. But the staff looked at me as if I was trespassing. They said no, I would have to wait 20 minutes. So I went around the corner to the family run coffee place. They too were in the throes of pre-opening preparation. I made the same enquiry, but this time was met with a big smile. 'Sure! Come in.' I always go to them for coffee now.

The new culture is about winning relationships. Socialeaders seek to build up social capital with clients and customers, with coworkers and others in the industry. This social capital is earned through help, support and kindness. Through actions that brighten a person's day. No social capital is earned by avoiding people. No points for being a hermit.

I'm not saying that we all need to be super-outgoing-customer-service-saints all the time, but we certainly can't hide from that which needs to be fixed. When problems arise, they need to be responded to quickly. Even better, when we see the opportunity to prevent a small issue from escalating into a crisis, we need to get proactively involved.

Through social media we provide this service publicly, which means that everyone can see both our problems and our solutions. They can see the process of how we help other people. We are evaluated based on our actions. We always have been, except that now it's digital. Our reputations can be Googled.

Building 'Social Capital'

Since you are no longer going to use traditional 'persuasive' marketing tactics, the consumers will have to decide for themselves that they really want to know you. This can make a very big difference to the relationship that ensues. We still persuade, but through our actions with others. Perhaps, someday, this will be such common practice that no one will take much notice of it, but for now people are so genuinely impressed to come across someone who actually cares, they tend to tell all their friends about it.

These are the beginnings of relationships. Hopefully we don't just help someone once and never see them again. Hopefully they come back, and we can continue to build a relationship. This is how I think of social capital.

Social capital isn't a value in an account. It's not a value that algorithms can parse, at least not yet. It is an entirely human feeling. In our hearts, we know how much we can ask of other people, depending on how much good will we have built up with them. We know when we are asking for more than we should; and we know that we will have to repay that kindness soon, somehow.

Sometimes we overestimate how much we can ask. Sometimes we underestimate. Social capital is tricky. But you know what it is: just think of asking a friend for a favour. Will they do it? That is your sense of social capital.

We like keeping our social capital in balance. Reciprocation is common because it keeps relationships on an even keel. I see this even on the micro-scale that is a retweet (passing along someone else's tweet with their name on it). When I retweet someone, they often retweet me back. I don't expect them to. I didn't ask them to. But they simply want to repay the favour. And in doing so, our social capital grows.

But social capital can also diminish. It depreciates quickly if left untended.

As so many social media exchanges take place publicly, we can see social capital between people online. We can see those who influence each other; we can recognize they have a friendship. While they may not be 'friends' in real life (they may not even live in the same time zone), the community can see that they share a bond. But it's difficult to appraise just what it's worth, for they may value their relationship differently. Social capital isn't some score we can neatly measure and track.

But the more social capital you have with those who have large shares of the master narrative, the larger your share will be too. This relationship requires service. Social capital isn't given for nothing. It may not come at all, or for a long time. Some people just won't be that in to you. Move on. Don't be too over the top. Woo with class.

Pro-Action is Better than Re-Action

Social media isn't a magic bullet that can fix everything that's wrong with your business, but they certainly can alert you to the problems.

Mistakes get made. It's how we respond to our mistakes that matter. It's got to be quick. In print journalism, a retraction was supposed to be more prominent than the error that it was correcting. It's important to make things right, and stand corrected.

This is happening a lot more with brands these days, but only after certain brands have already failed. And it is happening publicly, which is great. You too can become one of the thousands of case studies of brands that have publicly gone out of their way to make things right.

But you could do even better. According to an Altimeter study,[2] 76 per cent of social media crises could have been avoided with proper training and a better response process in place. There are a lot of opportunities in social media. The greatest are those where you can prevent a small issue from mushrooming into a full-blown crisis.

This requires being alert. But we can't be there 24/7. Our community, on the other hand, is there at all hours. Having lots of great relationships is a bit like having a neighbourhood watch – others can alert you to potential threats.

The granddaddy of all these is Scott Monty's TheRanger Station.com case.[3] Scott Monty is a brandividual pioneer working as Ford's head of social media. The Ranger Station is a Ford fansite. Fans matter a lot to any brand, but sometimes they get a bit rowdy. The Ranger Station was selling a Ford emblem that had the silhouette of a slender, busty naked woman draped over it. The Ford lawyers were threatening to sue. The big automotive blogs caught wind of this and started to scream that Ford was a big meanie for suing fans. Monty contacted all involved and managed to get the lawyers to back down and the fans to agree to become part of the official brand licence programme. Crisis averted thanks to a quick response.

Had Monty not seen the reaction of the automotive industry blogs, he wouldn't have seen the chance to do something. The lawyers would no doubt have sued the fans, and those fans would have turned against Ford forever, damaging the brand far more significantly than their unauthorized Ford emblem ever did.

The actions of Scott Monty have become something of fabled legend, retold as social currency by those wanting to impart the importance of being proactive.

The lesson for Socialeaders today: don't tell us 'We care about fans' – *show* us you that you care.

Great Service is Always a Great Story

The other granddaddy of customer service case studies is that of Larry Eliason's work at Comcast.[4] He started the account @comcastcares. He got out from behind the brand, put his face on the account, and would respond to hundreds of complaints every day.

It became the stuff of legend: so many disgruntled customers, so many of them completely turned around by Eliason's tweets. It wasn't just that he responded, but that he responded in real time. The immediacy mattered. Being able to interpret the problem and deliver a personalized response mattered. He provided authentic caring that turned complaints into opportunities.

Those are the same five reasons I gave for why we can't automate or outsource our social media presence: immediacy, interpretation, personalization, authenticity and opportunity. Combine these with great service and in no time we are building social capital.

Make no mistake, these exchanges are social currency. They are part of our story. They do become part of the master narrative. If done really well they become legendary case studies.

Of course, many brands are emulating this now. And, sadly, we are faced with a wave of early social media adopters who wish to exploit our customer service kindness. These early adopters have discovered that they can threaten to tweet about how they were 'mistreated' and get complimentary goods and services to appease them. It's the equivalent of extortion.

All the more reason you need a large network of true supporters. You need to be able to check out the online influence score of the person who is trying to blackmail you into giving them an upgrade. Can they back up their threats? If not, turn to the community for support. See if those who are your loyal customers think the extortionist deserves better.

Keep as much as you can transparent. But remember that means you have to be squeaky clean. Keep your cool. If you have to deal with someone who is real nasty, and you just want to give them what they want to shut them up. Move the conversation to a private channel. But if you can solve people's problems the way @comcastcares did, publicly, it makes for a much better story.

Eliason has since moved on from Comcast. His new focus as Citi's senior vice-president of social media is 'building a lifetime of trust' with customers.[5] That is where the Social-eaders are setting the bar: lifetime of trust. Now that's a legendary story in the making.

Take Away Express #9

- ☐ Social capital isn't a bank value. It's a personal measure of a relationship. It's a collection of favours and friendship, help, support and kindness within you community and between individuals.

☐ It's what you do, the choices you make, that matter more than the exact words:

- fix mistakes fast
- see opportunities to prevent mole hills from becoming mountains.

☐ Publicly providing great service is social currency that others trade. Solve many people's problems at once.

10. FRIEND RELATIONSHIP MANAGEMENT

'All animals are equal, but some animals are more equal than others.'

– **George Orwell, *Animal Farm*[1]**

As we transitioned from the mass media to the social media era, we brought some old evaluation techniques with us. Success in the early version of the web was measured by how much 'traffic' you could get. It was all about volume. And it didn't take long before we could claim millions of 'hits' to our sites. It also didn't take long to realize that that metric was pretty flimsy. So we tried coming up with something more authentic – like counting the number of unique visitors. But even measuring unique visitors has become grossly myopic.

Not every visitor is created equal. They are indeed unique. That was the lesson from 'The Chewbacca Incident'.[2] Michael Heilemann wrote an amazing blog post dispelling many of the myths surrounding *Star Wars* and titled it 'George Lucas Stole Chewbacca. But It's Okay.' *Star Wars'* fans streamed in from all corners of the web as the article was picked up and shared as hot social currency by all the major geek blogs in the geek narrative. While the article is a fantastic illumination of the creative process involved in developing a mythology as rich as that of *Star Wars*, Heilemann's follow up study of the visitors to his article, 'The Chewbacca Incident', provides us with the true illumination for identifying quality relationships instead of quantity of relationships.

Heilemann discovered that readers that came from different communities spent different amounts of time on the site. These communities had formed around various geek blogs. Daring Fireball, Hacker News and Boing Boing blog communities stuck around the longest, showing the most engagement. Visitors coming from Facebook spent more time on the site than those coming from Twitter, who simply looked for the punch line, retweeted and moved on. Reddit, where the social currency traded well over a longer period of time, sent a steady stream of visitors for a week. His analytics revealed that some communities are more engaged than others, and dispelled the myth that all traffic is equal. When we look at traffic this is the sort of referral information that can help us understand where to find our most engaged fans.

We could go a step further and look within each of those communities to find more layers of influencers. Just as not every community is equal, not every member of every community is equal. Using tools like Facebook connect, we can let fans opt-in to show that they were on our blogs and liked it. We can then evaluate each of these fans to see who has the greatest reach and influence.

Analytics may not be the most glamorous part of social media, but it is essential. It allows us to experiment and measure the results. But we must learn to measure the meaningful interactions, the social currency exchanges that develop social currency. We need to measure this one person at a time.

The Quantity Fallacy

It's been said that there are two types of people in social media: those who say they want more followers, and those who are lying. We can't help but want to be at least a little popular. It's long been connected to success in the mass media era. But followers, like traffic, are not what create influence. Having loads of subscribers does not necessarily mean you have community.

It's so easy to subscribe or otherwise befriend and follow a person without ever really checking out what they are doing. Likewise, it's very easy to get thousands of bots to follow you. None of them is going to engage with you on subjects that matter. You can be sharing social currency gold, but, even if there are thousands of them, they won't care.

It's possible to generate social currency so well targeted to keyword searches that it attracts a high volume of visits, but creates no community. This is the province of content farms, businesses that churn out shallow little answers to the most popular keyword searches online. Through SEO tricks, they managed to get their flimsy social currency to the top of Google's search. It's really just link bait. While they achieve the goal of getting traffic, and picking up that tiny fraction of a percentage of people who subsequently click on one of their ads, their social currency seldom solves our problems, which makes it worthless. Which means our relationship with them is also not positive. It makes us feel like we got burned when we click on the top result and get rubbish. How many people must have been burned by this?

Measuring visits and clicks doesn't reflect the complexity of relationships. Not all 'friends' respond to your cries for help. Not all family members get along. Some co-workers are best friends, other bitter rivals. Some friends live far away, but have a place that is always close to our heart. Neighbours who live next door may be complete strangers. Ultimately, with relationships, it's always 'complicated'.

Google's priority Gmail algorithms understand that some conversations are more important than others. So does Facebook's news stream. So do many of the newsreaders that tap into our social graph. These tools can be a big help, but they may force you to rethink, professionally, whom you are spending time with. The algorithms will make the echo chamber louder. You must seek to interact with the very best you can, not just with those you see every day.

There is a lot of noise out there, and everyone is getting better and better at filtering it out. How do you avoid being filtered out? What do you have to do to stand out, to get recognized and to get recommended by the right people? Which stories create the most value?

The new metric is that of influence. Like it or not, it will follow you around for the rest of your life. I'll go into online influence scores in more detail in Chapter 15. For now, know that these numbers, like Klout scores, are still in their infancy, and often subject to the same manipulation that content farms use to get top ranking in search. However, online influence scores do let us make some quick and fairly reliable judgements about who are the most important people to be seen with.

That feels snobbish to say. I'm not prescribing that you abandon friends, family and colleagues just so that you can be a social climber. But I am saying that you need to make some choices. There isn't enough time in the day to develop close relationships with everyone in your master narrative.

So the question is not, 'How do I get more followers?' but rather, 'How do I manage relationships with the most influential bunch that I can find?'

How Much Community Could a 'Community Manager' Manage if a 'Community Manager' Could Manage Communities?

The role of community manager is fast growing. The idea is that someone special should be managing this social media stuff. This reminds me of the way that big companies first had switchboard operators when telephones were being introduced into the workplace. Someone special was hired to manage the telephone stuff.

Eventually, we all just learned how to use the telephone ourselves. It just scaled a lot better that way. The same is true for communities.

Besides, your value as a professional requires you to have an awesome network of your own. Don't let someone else rob you of that. You career will rise on the success of your network.

This is where we need to return to the research of Dunbar, who gave us our Central Plank concept I mentioned in Chapter 1. Dunbar has found evidence that the human brain is only capable of managing about 150 relationships, something now commonly known as Dunbar's Number.[3]

Dunbar's Number comes as a result of primate evolutionary biology – we come from a long history of living in relatively small groups. Life in the giant cities is a relatively new thing compared to the human brain. Yet even in big cities, we have connections with a limited number of people, the rest we just ignore.

Whether our relationships are online or offline doesn't really matter – we have a pretty small number of people to choose to be 'friends' with.

It seems, by looking through my contact database, that we can certainly *know* a lot more people than 150. But I'm talking about managing relationships, which is harder than just remembering a name and a face (something which is hard enough).

This means we need to focus our efforts. Don't make it your objective to build up a community of thousands (at least not yet). Instead, we want to build a Dunbar Portfolio – a list of the top 150 relationships that we can manage to benefit our community.

However, if you work in the B2C world, you are probably wondering how this is going to scale. You still need scale. You need to sell lots and lots and lots. Don't worry, you will – at least if you make sure you are connected to those who are super-connectors.

Think about the network effect of being connected to a 100+ relevant people who are well connected to a 100+ more relevant people, who are connected likewise in turn. If you develop relationships that result in a high chance that each of these people would amplify and/or engage with the social currency you shared, then you would have a much bigger, much more engaged audience than you would achieve by any mass marketing means.

Just as the role of the switchboard operator could not scale to keep up with all the phone calls of a fast growing business with hundreds of employees, so it's impossible for the role of the community manager to manage all the relationships a business has with hundreds of employees. The employees need to manage those social media relationships, and the Socialeader must lead by example.

Finding Community Influencers

While we may think we are going to go out there and 'build' communities, the truth is that they already exist. Just because you are new to the scene, doesn't make the scene new. Instead of reinventing the wheel, look for pre-formed communities that you can make your own.

But don't think you can just show up and take over – even if you've got loads of money and good intentions. Brands have been known to try to 'co-opt' cultures to help make themselves look cool. They tend to be rejected outright by those with the most 'authentic' influence in the communities. You can't co-opt a culture.

That's not to say that some brands haven't been able to get seriously connected to sub-cultures. But it's been through hard work. It's about developing trust, showing respect and paying dues. It's not about getting one person to use a product and service. Brands like Red Bull have proven to be consistent adrenalin junkies over time, not all at once.[4] You can't fake it, you can't create a community overnight.

Red Bull has used a combination of sponsorship and word-of-mouth marketing, combined with extensive creation of social media and branded content effectively to win over their community.[5]

It's about embracing a culture and empowering it: true co-operation. Get involved and add value. You don't have to be an adrenaline junkie. Communities come in many flavours of cool. You can find them around places known for creating influence: fashion weeks, tech conferences, music festivals, business summits. Look at any event based on passion and you will find influencers. And usually, not far behind, some brands that are trying to get in on the action.

You need to earn admission to the club. Start by mapping all the digital influencers active in the community. Know who it is you need to court.

Also, remember not to dismiss people just because their online influence isn't up to scratch. There are still also many people in the world who are influential offline, but have no online clout.

Look to leaderboards: directories like WeFollow.com or Klout Topic Pages. They are not always accurate, but will provide a list of the 'usual suspects'. Begin your courting here.

Courting Royals

In every community there is royalty: those who occupy the short head (as opposed to the rest of us in the long tail). Royalty deal with a lot of noise. They use strong filters. Those filters will keep you out, unless you can find a way to impress the community.

Start with the basics: comment on their blogs, provide valuable insights, share links to relevant and cool social currency. Source and share with them the most valuable stories, insights and information you can. Don't worry if they don't respond at first. Consider sharing social currency targeted at tickling their fancy in places that the royals frequently visit, places that they care about. Sometimes they hang out in groups or chats – the cocktail party conversations of social media.

Get involved in these cocktail party conversations, minding your etiquette of course.

Instead of promoting yourself, take time to promote whatever it is the community's big influencers are promoting. Don't be sycophantic about it. If this is truly your community too, you shouldn't find it hard to be genuinely enthusiastic about the stuff that is happening.

Influencers always need help. As do the communities they love and that love them. Help out in a significant way. The more you give, the more they will give back. This can be something as simple as making a useful introduction to a business colleague or something massively generous, like providing access to a resource you have at your disposal.

If you know someone who knows the royal you are courting, ask them to introduce you. The context in which we are presented makes a big difference to whether or not we will get past the filters.

Once you've gotten noticed, take the opportunity to properly introduce yourself. Start by saying 'hello'. Find out the best way to connect one-on-one. Some channels are just too noisy. Most people have preferred means, be it email, telephone calls or, if you are in the same city, face-to-face meetings.

As you gain trust, ask for small commitments. Gauge your social currency carefully so that you don't ask too much. Once you receive, then give some more. Keep the cycle going. Be likeable throughout and make the experience as enjoyable for as many people as possible.

Courting royalty takes time and, like all online relationships, requires careful management. But once on the other side of the velvet rope, your influence within the community will grow much more quickly.

Take Away Express #10

- ☐ Not all 'traffic' is equal. In the mass media era, we wanted volume. So did the first version of the web. So did we at first in social media. We know now that a few key community influencers matter much more than thousands of bots.
- ☐ As you build your community, sort and organize. People can be in more than one list. Relationships are often 'complicated'.
- ☐ There are physical limits to how many relationships we can actively manage. Who's on your Top 100 list? Map out all those with influence in your community.
- ☐ Reaching influencers is harder, as they deal with increased noise levels by using stronger filters. You need to get on their radars.

11. FRIENDS IN NEED

'It's the friends you can call up at 4 am that matter.'

– Marlene Dietrich[1]

I often get cornered at networking events or conferences by someone I don't know who wants me to tell them what the ROI will be on social media for their business. They are looking for excuses not to get into social media, so they demand proof from the outset that the return they are going to get on all this effort will be worthwhile. This is never an easy question to answer. Without knowing anything about that person's business, I can't exactly look him in the eye and say, 'Your sales will be up 20 per cent next year.' I'm sure that's what he'd like me to say. But that would be to undervalue his return by reducing it to purely a bottom line endeavour. The truth is that the value we extract from social media might be better than that. It might be something money can't buy.

Like many investments, building strong relationships with a community isn't about the short term. Sure, we can make social media about how many sales conversions we can generate per tweet. There is no reason we couldn't track such things and make them key performance indicators if we want to. Just as long as we are also thinking about the long term, and saving for a rainy day.

Social media have brought new opportunities and new risks. Social media tools have made collective action by the masses easier than ever. While that action has been credited with playing an important part in unseating more than a few dictators, it is also the source of much crisis management among brands.

It is now quite easy for an angry horde to show up at your digital door brandishing the social media equivalent of pitchforks and torches. How do you plan to control the community once it turns against you? Social media crisis management is a crucial element for every business these days, and it begins with long-term investment. When the time comes, you will need your loyal fans and influential advocates to stand by you.

This means we also need to start thinking differently about what loyalty is. It isn't about the loyalty programme points. It's about receiving ever better experiences from each other as our social capital deepens.

Breeding Loyalty

Often the efforts we first think of putting into social media are those that build towards a sale. Maybe this is because the marketing departments have been the early adopters of the new tools. The pressure is there to use social media to make money. The urge to use it to raise awareness, interest and desire for a product or service is strong. But that urge should be no stronger than the urge to build post-sales loyalty. After all, that is what community is really all about.

Loyalty programmes are nothing new, but they are undergoing a transformation in the social media era. In the US, out of $48 billion in traditional loyalty points issued every year, nearly $16 billion remain unused.[2] If we are all so loyal, why does so much value go unclaimed? The problem is a misconception about what 'loyalty' is here. People don't feel loyalty towards each other because of points. Points only create loyalty to programmes. Worse, points that can only be redeemed for discounts focus the community's values on the customer's money, not on their loyalty or advocacy. Essentially points are just another way to try to buy a community.

In what story does the hero ever buy loyalty? Buying henchmen is what the villain does and, in a crisis, it's always his undoing.

Efforts are being made to move away from the usual travel-related redemption mentality. American Express created a rewards programme campaign called 'Social Currency'. The idea was that American Express customers who were also Facebook users could redeem points online for a wide range of interesting experiences (like concert tickets or downloads to their mobile phones) that had nothing to do with the usual sort of loyalty programmes. Because these experiences are 'cool', American Express hopes that customers will then share these experiences via Facebook as 'social currency'.

This is a step in the right direction. If that social currency then deepens the bonds among the members of American Express's community then it's on its way to building loyalty. But if their so-called social currency isn't the sort of social currency that's valuable to others, then the plan risks falling back into the idea that we can buy loyalty.

Ultimately, we don't need reward points, but we do need a system which ensures that those we exchange social currency (or monetary currency) with feel they are cared for and important members of the community. And to provide an experience that makes someone feel cared for means you need to really know your community. We need to use social media to find out more about the people who matter to our business, our industry and our master narrative.

When we sit down in front of our social media suites, we need to make sure that friends who are showing up in the streams are being looked after. Do we see them having problems? Solve them. Do we see them talking about purchases they've made from you? Make sure they always get the VIP treatment. That can be as simple as responding to their tweets or as complex as connecting them with a once-in-a-lifetime experience.

You'll be able to recognize their loyalty in return through the passion they express for your brand. Then when it comes time to talk ROI, you can verify the studies that show there's a greater likelihood that those who talk about you online will pay more to remain loyal to your brand.[3]

You'll also be able to call for supporters when you need some help.

Social Capital in Times of Crisis

We live in times where social media has made everything public. The tools of participation are available to nearly everyone. It has never been easier for people to rally around a cause or calamity. The detractors may come for you, and it may be wholly unfair and unwarranted. How do you quell the angry mob?

One of the methods is with an outpouring of love from your most influential supporters, those with whom you've built up a positive balance of social capital over a long period of time. This is the return you want to get from your social media efforts.

Their authentic voices – singing your praises – are a far better thing than any paid for endorsement.

Social capital is the trust we have earned through our actions and words, our tweets, our likes, our comments and online exchanges of social currency. It is a reflection of our personal and professional interactions. But how much are we going to have to give in order to get back?

Ask yourself, how much would you need a brand to invest in you before you'd be willing to defend it? For fans of Apple, the company does very little besides provide excellent products and great customer service. Yet Apple fans are some of the most vocal advocates you will find online. They are willing to engage in fiercely loyal battles with tweets, posts in online groups on LinkedIn or on Facebook Pages to defend the awesomeness of their consumer devices.

We can't all be like Apple. For most, advocates are born from community relationships where the interpersonal threads weave us together into our narrative. The struggle of good people to overcome attacks by evil trolls is one of the great social media conflicts we can all act as heroes in.

When you call for help for someone else, it is much more powerful than when they call for themselves. It's also much more touching – emotionally charged – social currency. When fans see tweets with people expressing their love for you even in your darkest of hours, it's really easy for other fans to hit the retweet button.

On Facebook, we may get one negative comment, but the defenders step in to reply on our behalf and, because of the way that Edge Rank works (the algorithms that determine what shows up in your Facebook stream), closely connected members of the same community will begin to see that friends are replying to a negative commenter in an attempt to right a wrong. Other community members then find it easier to follow suit, and add their own positive endorsements. In response to just one hater, a community of positive comments begins to pile up to the point that the social proof of love outweighs the hate. 'Hundreds of lovers must be more right that one detractor', is how we think about these things.

Negative attacks on the likes of Justin Bieber stand little chance in online arenas. Be it Facebook or Twitter, Bieber's army of adoring teenage fans will defend their beloved star until the end (or at least until their mums tell them it's time for bed).

What Needs a Response (or How to Kill a Troll)

Stand by friends in a time of need and, so the theory goes, they'll stand by you. But not every negative attack requires action. Sometimes it requires ignoring the problem.

Now, I know that doesn't sound like the sort of proactive advice of a Socialeader, but it is sometimes the best advice when trying to kill trolls. Trolls, also known in some circles as griefers, derive enjoyment from irritating and annoying others online. They feed off the negative attention. The trolls in fairy tales often live under bridges. They jump out and cause mayhem for the unsuspecting heroes. The trolls we find online are not dissimilar.

Trolls usually hide behind secret identities or anonymity. This is one of the major reasons you'll find so much social media moving away from secret identities. In blog comments that have required people to log in with their authentic Facebook profiles, the number of trolls has dropped drastically. The reason for this is simple: it's harder to be an obnoxious jerk if the world knows your name. After all, few want the online reputation as an obnoxious jerk for the rest of their lives.

But sometimes even real people can be mean and critical. The worst are the attacks that are purely personal. At some point online, you will face a spiteful, personal attack. As an individual serving a brand, there is no logo to hide behind, and the attack can hurt. Sometimes the response to this is for us just to develop some thick skin – a key trait of a Socialeader. If you've never experienced this before, it will shake you. Your emotions will explode. You'll want to respond with guns blazing. Don't. Wait. Step away from the keyboard and breathe. Try to laugh it off if you can (the more times you face this situation, the easier it gets).

Ask some friends in the community for some perspective. Is there anything about the attack that is true? Do you have anything to really worry about? They'll likely tell you to ignore it. Good advice. But I'll bet that it still sticks in your head for a few days.

There's little point in responding to every attack. Instead, if the personal attack is unjustified, and they usually are (they usually have everything to do with the jealousy or insecurity of the attacker), simply use the built-in social media tools to block the person. If they can't be civil towards you, or you can't find a way even to respond to them without losing your cool, then there is no place for them in your community. Blocking them, of course, won't stop them from libelling you, but in the end they often do more damage to their own reputations by making unfounded personal attacks on someone. Ultimately if they continue long enough, and you can prove that their malice is having a negative effect on your business, livelihood or reputation, you can seek legal action. Libel laws are there to protect you, but hopefully it won't come to that.

If you see someone you know getting slammed online in a personal manner that you know has to hurt, be a friend, stop by and tell them to ignore the hater. Repair the damage with positive encouragement. We can help each other develop the thick skin we need to survive online.

But don't ignore legitimate complaints, even if they are made by someone in anger. Often if a customer is really angry and you can fix their problem, they become a super loyal supporter. Customers in need require great customer service. But never feed the trolls.

A Cautionary Tale

Be careful how you wield the vengeful spirit of your community. Asking your fans to defend you can be a dangerous tactic, transforming them into the online equivalent of a lynch mob.

A bartender named Victoria Liss, working in Seattle in 2011, received a very rude message scribbled at the bottom of the credit card payment slip: 'P.S. You could stand to lose a few pounds.'[4]

Along with the insult, the customer marked a big zero for the tip.

Outraged, Liss turned to her sizeable friend base on Facebook. She posted a picture of the receipt with the message written on it. She called him 'yuppie scum.' Her friends agreed and decided to go hunt for the offender to let him know. The name on the payment slip read 'Andrew Meyer'.

This hunt then picked up additional traction from local blogs with sizeable communities. Outrage makes for powerful social currency, and it wasn't long before it had been passed along to the big league blogs like Jezebel, NPR and Gawker, who brought it even greater attention. The story got so much attention that it even crossed the pond with the *Daily Mail*[5] reporting on it in the UK. People around the world were now outraged, and ready to teach this guy a lesson. They scoured the web for this man. And then someone found him. Or rather they found *an* 'Andrew Meyer'.

But they got the wrong guy. It didn't matter. The mob was on him. After 68 spiteful messages, the wrongly accused was considering legal action.

Liss attempted to intervene. She was deeply apologetic and tried to quell the angry mob, but it had gotten far beyond her control.

As a Socialeader, you need to show more restraint. Outrage does make for powerful social currency, but it is terribly negative and brings little benefit to any of the parties involved. Before you decide to retaliate, ask yourself if it is really worth it – and make sure you've got the right person.

Take Away Express #11

- ☐ Loyalty isn't about points. Know what is important to your community and find ways to make their experience even better. Stand by them in moments of need – that's true loyalty.
- ☐ Don't cash out too soon: the return on investment for social media may not be immediate, but it may come in a form that money can't buy. When you are in crisis, you'll want some social capital to draw on.
- ☐ Defend others as you would have them defend you. The sort of thick skin you need to be a Socialeader takes a while to develop. Offer support to those who have been attacked online.
- ☐ Don't respond to trolls. They feed off the negative attention. Learn how to block those who make personal attacks.
- ☐ Great customer service can turn angry customers into loyal supporters.

12. DON'T BE A SOCIAL MEDIA SNOB

'The true definition of a snob is one who craves for what separates men rather than for what unites them.'

– John Buchan

Don't let yourself go Hollywood. Don't let all the bright lights of becoming a minor weblebrity blind you to the little people who helped you rise to the top. While it's no doubt important to be investing time in super-influencers, they represent only the core of your online presence, not the whole. You need to be bigger than that, which means you need to be looking out for the little guys too.

I've often taken criticism for following back so many people on Twitter (more than 80,000). Some have claimed this makes me disingenuous, saying that if I truly cared about those I follow then I would read everything they tweet, something I can't possibly do because I follow so many. It's true, I don't read every tweet that every follower writes. I don't think anybody with more than a few followers does.

I look at it differently. I've tried to follow back everyone that follows me in the same way that I try to shake every hand that gets extended to me at a conference. We shake hands, maybe we talk for a couple minutes, but there is no obligation that we do business just because we were decent enough to introduce ourselves. But I will be much more open to receiving an email or a private direct message on Twitter from you (which, by the way, is only possible if you follow someone, and another good reason to do so).

Mutually following each other is simply a pleasantry that I appreciate together with members of my community. I know that many of them have appreciated the courtesy too. There is nothing disingenuous about it. Actually, I think the opposite is true: not being open to connecting to new relationships is tantamount to snobbery. You're not so important that you can't look out for the little guys.

Those who think I follow too many people on Twitter should also know that I probably subscribe to too many blogs too (nearly a thousand). I'm also now subscribing to too many pages and profiles on Facebook (hundreds). I've probably got too many connections on LinkedIn (5,000+) as well. Well, too many for some. I know that from this crazy mass of connections comes unexpected value. Social media tools make it easy to filter and search through the noise. The more connections I have, the more chances I have that someone might provide unexpected value.

By closing yourself off to others you close yourself off to the value of the community. Sometimes the little guys can surprise you. Little guys can become big guys. Or those who don't appear terribly well connected online can be very well connected offline.

So while we spend the bulk of our time dedicated to strengthening the relationships among the few, we also need to stay relevant among the many. A true star always makes time for fans. Nobody likes a social media snob. And if they don't like you, they won't listen to you (even if they still follow you back).

One Click Away from a Superfan

Loyalty comes at all levels and is often more profound than you expect. Maybe you've shared a few things that really touched someone, then one day you shared something so emotionally charged for them that they look up to you forever more. Maybe as you've grown in stature they have grown to be even bigger admirers.

It's difficult to know which click of your mouse distributed the social currency that tipped someone over the edge. But the more you share, and the more you engage with people, the more likely a small but impassioned group of followers will emerge – especially on very open social media services like Twitter. These are the superfans, and they don't need to be influential to provide you with serious support.

Wired co-founder Kevin Kelly's concept of an artist needing only 1,000 true fans to survive[2] describes superfans as the sort of people who will 'purchase anything and everything you produce'. The idea is that these fans are so fanatical that they will supply an artist with $100 each – a total of $100,000. For most artists, that's a living.

Even if you're not an artist in the traditional sense, and your superfans aren't quite ready to drop a $100 on you, you will recognize the important fans as the ones who re-share everything you share. They send you a disproportionate amount of fan mail. They push lots of non-content social currency towards you.

As awkward as this might feel, don't discourage them. Instead, consider investing time in your superfans. Give them the VIP treatment. Help connect them with other influencers in the community. As you grow their influence, yours grows too.

Superfans are loyal advocates who, if you treat them right, will be there to defend you in a crisis. And there is a strong chance that they will be the ones who buy your new products. Once you identify superfans, you might even consider sending them products to review. This isn't astroturfing: you're not artificially buying positive reviews. These are people who will fill Amazon or TripAdvisor (or wherever your products or service get reviewed) with the five-star passionate reviews that other people will believe. We must seek ways to turn superfan enthusiasm into emotionally charged social currency that we can put on our Facebook Pages or re-share on Twitter.

And it's not just the superfans who matter. There are lots of other 'little' accounts in social media, held by very 'big' people. One of the purposes of this book is to help people bridge the gap between their online and offline influence, so we must remember that there will always be those who don't have a large network or a major share in the master narrative, but who do hold much more non-digital value. Don't underestimate offline influence connected through small online encounters. Any small fan in your community may be a though-leader in their real-world community.

Of course, none of this means you are under any obligation to accept from superfans their offers of marriage or to have your babies. The longer you spend online, the more likely this will happen to you. It's the opposite of being personally attacked, which I discussed in the previous chapter, but it's usually just as unwelcome. I believe that, loyalty or not, there are lines you shouldn't cross. What I've been suggesting we do with social media is make it more professional, and less personal. But matters of the heart know no bounds.

Regardless of where your heart may lie, take any such personal conversation into a private channel. On Twitter use DMs. In Facebook, send a private message. No fan should get their heart broken in public (be gentle, but firm).

Celebrate the Little Guy

Remember @theashes, the young American woman whisked off to the cricket tournament in Australia (see Chapter 3). She wasn't an influencer, but she became a big part of the story. She became influential overnight almost by accident.

The creation of a superfan can happen unexpectedly. It may have everything to do with seizing an opportunity just as Qantas and Vodafone did with @theashes. You must be ready to help the little guys become heroes. It's a crucial part of building community. While not every one of us can become a hero, we can live vicariously through a peer who does get a chance to stand in the spotlight.

Sharing in someone else's amazing experience can make for very valuable social currency. This has been the basic formula behind television talent shows for much of the past decade. We talk about the singers as week by week they win their way towards minor stardom. Their performances become the social currency we exchange around the water cooler. But when it's all over, we stop talking about them. The winners do their best to exploit the heightened public awareness of their existence, but the value of their social currency usually diminishes quickly. Those who made the stars, the judges and creators of the talent shows, retain far more influence from the experience.

The influencers then once again open the doors to the widest range of 'little guys' they can and start looking for opportunities. Social media, while not nearly so cut throat and exploitive, is much the same.

Show the community that you are accessible to all. Play the part of the talent scout looking for rising stars to celebrate. Treat them with respect and admiration. That's one of the great promises of social media: respect and admiration get paid forward.

I've fallen out of my chair several times when giant heroes of mine have mentioned me in a tweet (Stephen Fry, Chris Anderson, Jack Dorsey). The feeling is fantastic. Pay it forward.

Also, make time to do the equivalent of signing autographs. It's one of those things that truly great celebrities always say to make time to do. As you rise in social media, don't forget to make time to thank people for sharing your social currency. Make time to get involved in a little #followfriday action (a tradition of thanking fans on Twitter, happens on Fridays, often marked with the hashtag #ff) or in any other means of spotlighting the names of the mere mortals who helped get you where you wanted to go. You need to discover how good it feels to get retweeted by someone bigger than you. After that, make others feel the same way and you will build a lot of social capital.

Give Your Quiet Fans Something to Do

More people are watchers of social media than they are contributors. Many will sit and read everything you share, but they will seldom get up and get involved. While they are reluctant proactively to share social currency, they may be inclined to have a little interactive fun.

This means that as a Socialeader you need to create opportunities for the shyer members of your community to get involved in some meaningful way. Think of ways to connect your true believers. The more interconnection that you create in the community the more valuable it becomes.

Ask them for help. Ask them for ideas. Do what you can to draw those lurking on the sidelines out onto the dance floor. On Twitter, I'll sometimes pass along news of a new tablet from a major manufacturer with the question, 'Would you trade your iPad for this?'

The shower of responses is never a surprise. Apple fans are vocal. They will shout 'No way!' To which I can respond with something inclusive like, 'No, I'm not ready to trade mine either.' And by doing so, I build a little social capital with some new friends.

But I'll also get others who will just offer me their opinions about tablets. And I can start conversations with these people, many of them people whom I've never spoken to before.

Drawing the real lurkers out of their shells may call for even greater measures. Once you've established a bit of presence, consider running a contest with a decent prize. News of this will make for powerful social currency. People will share it with friends across Twitter and Facebook very quickly. The prize doesn't need to be of great monetary value. Often people are far more interested in access. For example, you could arrange a meeting with someone influential. It could even be with yourself or someone you have access to.

The geek circle I hang out in love the ability to get to a 'private beta' test of a hot new service (it's a bit like getting into a hot new nightclub that hasn't even opened yet). Access to special or exclusive treatment is often enough of a prize, and it can be especially tailored to your community rather than to the general public. Everyone will show up for a free car, only those who are serious about your business will show up for a free webinar with the CEO.

Think about using this as an opportunity to strengthen the community. Ensure that the entries to the contest are valuable social currency that you can share. For example, when someone signs up for the contest, give them the option to tell all their friends on Facebook that they are in the running for access to something very exclusive. Make sure the invitations you give them to pass on to their friends are funny or emotionally charged. Get others in the community to experience the fear of missing out. This will spread the word far and wide, and draw out many who have never before been motivated enough to make direct contact with you.

If a contest seems too much effort at this stage, start smaller. A similar effect can be had through the creation of a funny quiz or other interactive toy that can be shared as social currency – and tracked.

Either way, the point isn't solely to create some sort of campaign to smoke out and capture all the small, quiet members of your community. Rather it is to keep them engaged with original social currency that they can't get anywhere else, social currency that binds them together and gets them talking.

Look for any opportunity to do this, from tweeting questions to throwing big contests. But make sure that you then personally reach out to everyone who responds on a one-to-one basis and make them feel included. If you don't know what else to say, just say thanks.

Don't underestimate the pleasantries, especially among the quiet ones in your community. Make some time to say 'Good morning, Twitter!' or 'Have a great weekend, Facebook!' You'll be surprised how many respond in kind. Step that up a notch and offer greetings to as many of your followers as you can. Watch how fast they are at replying to that sort of contact. People may be shy when it comes to weighing in on the big issues of the day, but it's easy to engage in small talk.

The bottom line: you can't afford to be a social media snob. While super-influencers will occupy the bulk of your online efforts, you also need to be tending to the party on the non-VIP side of the velvet rope.

There is so much value that can be found on that side of the velvet rope if you stop thinking about these people as passive followers. They are no longer just consumers or customers or fans. That is the mass media way of thinking of them. We need to embrace the social media way of thinking, the mom and pop shop way of thinking. The people formerly known as 'the audience' are all now potential collaborators in the production of our story. They are a crowd from which value can be sourced.

We need to come down from our Ivory Towers and ask for help. Ask for advice. Be humble. Support others as you'd like them to support you. That's what Socialeaders do.

Take Away Express #12

- ☐ Nobody likes a social media snob.
- ☐ You never know when or where value may come from in your network.
- ☐ The bigger the network, the more opportunities you have.
- ☐ Online influence isn't everything. Sometimes people have a lot of offline influence. Sometimes they gain influence overnight.
- ☐ Keep the party going: give the quiet fans something meaningful to do.
- ☐ Pay it forward: shout out the little guys the way that you wish the big guys would shout out your name.

13. TAKE IT OUTSIDE

'Not all those who wander are lost.'

– J.R.R. Tolkien[1]

I often get tweets from people whom I've met on Twitter who are passing through town asking if I'm interested in meeting for a coffee. I nearly always say yes as it's often a wonderful opportunity to get to know the real person behind a familiar online presence.

It can be both exciting and sometimes disappointing. I often joke that people look taller than they do in their social media pictures. What I really mean is that they look nothing like their social media picture!

This is the reaction you can expect to get from people if you choose to use some other online form of identity than your own. If you don't have a real name or real picture, then you have to introduce yourself to people by some ridiculous Twitter handle. I have people come up to me all the time and say I'm so and so on Twitter, a handle I immediately recognize from our many exchanges, but I'm left with a strangely disconnected feeling like I really don't know this person at all. Inevitably I forget their real name almost at once, while I continue to have to call them by their Twitter handle throughout the rest of our meetup. It's awkward, to say the least.

It's through these real-life encounters with people I've met online that I can truly sense the gap between the perceived intimacy of social media and the offline realities. Some argue that social media has lessened the social ties between communities, replacing them with an online substitute that at best offers weak ties rather than strong bonds.[2]

For the vast majority of human life on this planet, there was no technology to convey social currency. Its exchange was limited to face-to-face meetings between individuals. Technology has continually eroded those limits, but has it come at the expense of real-life social capital? In fact, if anything, the technology, from the postal service to the telegraph to telephones to email to social media, has made it easier and easier to coordinate offline meetings.

Social media's most surprising effects have not been played out online, but outside in the streets. When Twitter was first released in the wild at the South by South West Conference, it had an effect that surprised its creators.

Jack Dorsey, Twitter co-founder, had written Twitter as a simple tool for updating an online status via SMS. What made it different from regular SMS was that you could subscribe (follow) others and get a text message whenever they updated their status.

But immediately users of the new service at the conference realized they could do something more with it. They could exchange social currency. It allowed people spontaneously to form something they called at the time 'Twittermobs'. Suddenly, to the Twitter founders' surprise, people would move in groups, abruptly leaving a bar based on a recommendation of something better down the street. People were swarming in real time like never before at the conference.[3]

Social media, along with mobile technology, were suddenly a real-world tool, not just some sort of online activity. Since then they have been used to fill streets around the world with protesters and rioters, with parades and cleanup crews.[4]

But you don't have to be overthrowing a dictator for social media to prove useful for your communities. Gatherings, often called meetups or tweetups, really do create stronger bonds between members of a community.

Of course, it's not always possible to bring everyone in your online community together in one place. Some key influencers may be based on the other side of the world. But people do travel a lot. If you live in a major world city, lots of influencers will pass through your fair metropolis.

If you live somewhere remote, perhaps you will need to get out for a tour of some of the social media capitals. Tell your communities when you are travelling in their parts of the world and maybe they'll throw a meetup for you.

Just make sure that you look like your profile picture.

Boots on the Ground

We still need real world contact with people. Even the best telepresence tools are no match for the real thing. Body language is a big part of communications. So are handshakes and hugs.

Socialeaders make time (and find budget) for conferences that are relevant to their business. This can be a crucial investment in community building.

Conferences are increasingly social media empowered with tweetwalls and apps to keep you updated on everything that's going on. These tools also provide a back channel for discussing the action on stage – a high tech means of passing notes.

If you use Twitter for this sort of thing, make sure you use hashtags to help everyone see that you are talking about an event that you are at. Increasingly attendees use hashtags, like #whatever2012, on all their tweets so that other attendees can more easily find and interact with them.

Track the hashtag by using Twitter search to set up a new column on your smartphone social media app (like Hootsuite or Tweetdeck) to stay in the loop with all the back channel chatterers. See who is talking. Check out their online profiles so you have a better understanding of who your fellow social media enabled conference goers are. Then invite people you'd like to meet for a coffee during the break so you can put stronger bonds in place.

Using hashtags while at a conference can also gain you more followers. If you can tweet lots of relevant insights and quotes from those speaking on the stage, many people who couldn't make it to the conference will follow you more closely to find out what they are missing.

But conferences aren't all about sitting in the audience. As a Socialeader it's possible to contact event organizers ahead of time and volunteer your expertise for a panel or even a keynote. Public speaking is a great way to help establish your authority with your community. Make sure to get your presentations up online for everyone to share as social currency. Also upload to Facebook any photos of yourself standing on stage – this is social proof that you are an expert, which we will see later is key to building influence.

Also, if you're up for a bit more work, try being a moderator on a topic that you know about. Rather than doing all the talking, this lets you be more of a community builder. This can be an opportunity to take questions from your social media community, ask panelists, then share their answers back to the community.

It doesn't have to be big conferences to make it worth getting out of the office and meeting people in real life. Small networking events can be just as good. Often there is a guest speaker whose nuggets of wisdom you can package as social currency and tweet or Facebook them out to your followers from your smartphone. Networking events can be great opportunities to meet new people who aren't yet following you on social media and invite them into your community.

And if you are carefully monitoring your master narrative, occasionally a spontaneous opportunity will present itself. It could be a flash mob or a demonstration or an exclusive party. Go. By being in attendance, you are more clearly viewed as an 'active' member of the community and given greater status.

Part of being a Socialeader is having an offline presence at the events that matter. Real world contact makes such an important connection in ways that joining online groups that share links will never amount to.

Checking in

Social media isn't just stuck on the web. It is very mobile. Through the use of apps on smartphones, you can use check-ins as a form of social currency that can bring your community together.

If you are attending conferences, networking events or even flash mobs, the thing to do is to check in on a location-based social media service. These range from stand alone services like Foursquare to integrated 'places' services with major sites like Facebook and Google.

Checking in uses the GPS in your phone to let you indicate that you are at a venue or event. These apps will show you who else are also checked in, which is a great shortcut for finding other people to talk to.

You can use them outside of events too. As an everyday tool, they can be used to let friends in your community know that you are at their stores or places of business. Check-ins to businesses demonstrate to your online community where your offline affiliations are located. They're a way of indicating physical presence in a virtual manner.

Also be aware that this can work against you. Don't check in just anywhere. Check-ins are social currency that need careful curation just like everything else you share. Checking in at the cheapest, nastiest locations isn't going to boost your online credibility.

But don't fake glamorous check-ins either. With many of these location services, it's possible to check in to venues that are a considerable distance away (an allowance they make as GPS has a significant margin of error). Keep it real. If you check in at an event, but don't go, then people at that event will be looking for you, sending you messages asking if you'd like to meet face-to-face. You'll be faced with having to make some awkward excuses.

Take Away Express #13

- ☐ Strong bonds come from meeting face-to-face. Our relationships become much deeper with our online community after we get a chance to meet them offline.
- ☐ Social media makes it easier both to identify opportunities to meet people and to spread the word about such opportunities.
- ☐ Meetups: use real-life opportunities to grow your community. Go to as many events as you can. Use the tools to coordinate face-to-face introductions. Make sure you look like your profile picture.
- ☐ Connect local community to travelling influencers.
- ☐ Use location-based check-ins as a form of social currency.

Part Three

INFLUENCE: PULLING BACK THE CURTAIN

14. WHAT IS INFLUENCE?

'If you get to thinking you're a person of some influence, try ordering somebody else's dog around.'

– Will Rogers

There is no simple answer to 'What is influence?' It has long been an area of intense academic debate, exacerbated now by the rapidly evolving social media landscape. Whatever influence is becoming in this new era, our futures are only going to need more of it, as it will follow us around and affect everything we want to do. Online influence is remapping our world.

When we go looking for online influence, sometimes we find it emanating from connectors and mavens, sometimes it sparks from average citizens. Sometimes it amalgamates with the content creators, while at other times it flows straight from the source, the great curators. We sometimes see online influence burst forth from leaders who rise from their communities to meet an unexpected opportunity. Sometimes we can barely feel the consistent stream of trusted advice that over the long term builds the unstoppable momentum of a leader. Influence is found in the heart of the little guys who stand up for their values. Influence is there among those who are ranked the highest on online leaderboards, even if their hearts aren't in it.

Some define influence as the ability to affect the choices made by others, with three possible outcomes: compliance, identification and internalization. Compliance is when you secretly disagree, yet publicly agree with the community to protect your status. Identification is when you want to be associated with popular influencers and so you align yourself somehow with their story. And finally, internalization is when you truly agree with influence both publicly and privately.

Influence can be used as a weapon: 'Do you know who I am?' While we've all had moments where we've lost our tempers with customer service representatives, we should never threaten to use our influence to bash someone into compliance. Violence, or the threat of violence, is not the weapon of influence that we want to wield.

There is another definition I like: influence is the power to convene and to tell a good story. Just getting people to gather around and listen in an era of wide-spread attention deficit is an accomplishment of influence.

Influence can be subtle. It can be lots of little nudges by a group to shepherd a new idea into popularity. As a group we can influence individuals much more easily than an individual can influence a group.

Influence is also personal and one-to-one. The closer your relationship with someone, the more influence you exert. But since we can only have a limited number of active relationships, there are only so many people we can influence this way. Every relationship is different. Everything depends on the context of the situation.

No single influence model seems to support all occasions.[1] But we don't need a grand unifying theory to be influencers. We have been influencing situations since we were babies.

When we wanted food, we would cry and food would arrive. Our methods have become more sophisticated over the course of our lives. And while much of it is subconscious, you will know influence when you find what works for you.

However, we can consciously seek ever more sophisticated and effective methods. We must be on the lookout for power techniques, both so we can use these to help move others and protect ourselves from their effects – we don't need to respond every time a big baby cries to get their way.

The Six Weapons of Influence

In his book *Influence: The Psychology of Persuasion*, Robert Cialdini established six weapons of influence: scarcity, authority, liking, reciprocation, commitment and consistency, and social proof.[2]

These six concepts have been embedded in everything I've been prescribing in this book. Through our use of social currency our exchanges and our building of social capital, we have been employing these so-called weapons. I'm not sure we should view them as weapons, unless you are using them in a purely exploitative fashion. Let's call them power tools. Anyone who has ever been on a construction site will agree that power tools can be dangerous in the wrong hands. But in the right hands they can build architectural marvels.

Often we are wholly unaware of these power tools of influence at work on us, but they are constantly present. As you read the news in your newsreader, and come across a blog post that is super relevant and only a few minutes old, you should feel a rush knowing that you might be the first one to share this valuable currency with your community. That's scarcity at work. While there is lots of content, there is only one person who gets to have the scoop, even if it only last for a few minutes.

The spoil of being first is that you, not your competitor, are the one who ends up with all the comments about the social currency growing underneath it on your Facebook profile. Timeliness is a scarcity – no one can make more time. Scarcity is a big part of the fear of missing out. We all develop this need to be kept in the loop.

And we want to be kept in the loop by someone of authority. Right from the beginning, we need to know what our own stories are. We need to take a stand as an expert on something, and to identify ourselves as such. Authority is a powerful tool of influence. We know that the public wants to talk to authority figures at brands.[3] We know that people want to identify themselves with authority figures – often they will follow you just because they think by following an expert it will make them an expert too. And we know that authority is often about titles and trappings – the professional bio and picture. It's important to have an online image that is carefully curated to make you look like an authority. We are more likely to follow instructions from someone dressed as a policeman than from someone dressed as a civilian – even if the uniform is a fake.[4]

Of course, it's easy to fake authority and influence online. Just because someone's bio says they are a Darth Vader, doesn't make it true. While we need to ensure our trappings are genuine and professional in appearance, we must not assume just because someone else has similar trapping that they are truly authorities. Once we know the tricks we can be more resilient in the face of those who are trying to trick us. As I've said before, everyone looks taller in real life than they do in their profile picture.

Naturally, I completely understand why people rarely look like their profile pictures. Doctoring your profile picture and online image to make you more attractive is an important part of the 'liking' power tool that Cialdini identified. Liking is fairly self-explanatory: it's about being likeable. While this starts with superficial image stuff, it's also deeper than that. When we act in a friendly manner, when we stand by our values, when we inject humour, when we are helpful – it all lends itself to making us likable. When we spew anger, hatred and resentment, we are difficult to like (hence why it's so important not to get drawn into online fights). We need to be the nicest, most giving versions of ourselves.

Giving things to people begins the act of reciprocation, which is another one of the power tools of influence. This is at work through almost all social currency exchanges and social capital building. Remember, people try to keep their social capital in balance. When someone pays us a small compliment, or offers us a greeting, we immediately repay them in kind. When someone needs a favour, we offer them help, knowing that what goes around comes around. Someday we may be in need of a favour, and those whom we have helped will be ready to level the social capital imbalance.

When we retweet someone, it increases the chance that they will retweet us. When we comment on something they posted to their Facebook timeline, it increases the chance that they will do the same for us. Most of the time, we don't even notice ourselves doing this, we act out of habit.

We are creatures of habit, and we like sticking to our habits – even to our detriment. The influence tools of commitment and consistency can be used that way, and have been by ruthless sales people for decades. This principle works along the lines that once someone makes a commitment to something, no matter how small, they are more likely to continue on that path.

For Socialeaders, we are employing commitment and consistency when we are building loyalty conversions, when we are asking for advocacy. By openly endorsing you, they are more likely to remain loyal to you. Facebook has used this ploy with their 'like' button – because it creates a public endorsement that you, an influencer, are putting your name beside this social object (even if you don't really like it that much). While we know that online commitments of this nature don't really produce strong bonds, the connections are not insignificant either. We will want to remain consistent in our actions, and not flip-flop or appear indecisive.

Finally, the sixth influence tool is social proof. This is probably the most clearly powerful of all in social media. This is what spurs on the crazy popularity contests where users want to attract the largest number of friends, followers, subscribers, connections, Klout scores or anything else that makes them look important. It works on the idea that if all these other people have done something (like followed you) then that must be the right thing to do. We will comply with the group, even if we secretly think that maybe the big number isn't all that important. Even our online influence scores, which are supposedly measuring quality over quantity, are becoming a more widely recognized form of social proof.

It's crucial not to get too drawn in by social proof. I know that is easier said than done. We need to realize that in the online world, these numbers can all be unfairly manipulated, and that they may not represent the true person behind them.

Socialeaders need to take the time to get to know people. Your own social skills, built up over a lifetime, will be a far better tool for measuring someone than any algorithm can produce. That said, we will be judged (or misjudged) based on these scores by those who can't be bothered to get to know us, so we best invest in growing them.

OIS: Your Social Credit Score

You walk into a popular restaurant without a reservation. You've never been there before. The maître d' asks for your name and types it into his tablet. Your online influence score pops up and indicates that you are a Socialeader with a big reach and lots of influence. The maître d' immediately moves a booking made a week earlier by someone with very little online influence to a spare table near the kitchen so that you can be seated like a VIP.

While that scenario is still fiction, that reality is coming. VIPs have always received special attention. If you were famous, or powerful, well connected or rich, you would enjoy a special level of attention, because there was an assumption that your level of influence could make or break someone.

Those with significant online influence scores (OIS) are now starting to enjoy the same VIP perks that used to be reserved for the rich, the famous and those who were connected to large media outlets. And for the same reason – perceived influence.

While OIS are far from perfect and subject to radical shifts as algorithms get updated and users learn to game them, they offer a major improvement on simply gauging someone based on how many friends or followers they have.

The most widely recognized of the OIS is Klout, a service founded by Joe Fernandez and Binh Tran in 2008. Their goal was to set a 'standard for influence'. And they are well on their way. At the time of writing, they were analyzing the exchange of 2.7 billion pieces of content, and the connections between more than 100 million people online every day. From this, Klout distils an online influence score that reflects your true reach, your ability to spread your social currency, and how many of the influential elite in your network interact with you.

How accurate is it? Well, influence is tricky. There are still nuances to be worked out. It's still possible to 'game' the system. Which means that, just as the importance of showing up high in Google's search engine created an industry of search engine optimization, there will no doubt be an OIS optimization industry, offering to squeeze out every extra point of Klout possible through tricks rather than hard work.

Despite those who may be trying to manipulate the system, the idea of having a single trustworthy influence score is very appealing to many. It's the sort of thing that has sparked a number of rival services like Peer Index, Kred and Twitalyzer.

Having a range of scores to compare is crucial. There is not yet one single means of measuring influence that can accurately portray everyone. There will no doubt be more OIS services joining the fray, each providing us with a different neatly decanted number that sums up all our online interactions and what we've inspired among others.

It's important to remember that, in these early days, OIS often measures only network activity and content amplification, not the movement of hearts and minds. It's possible to influence people in ways beyond getting them to comment or retweet you. So don't worry if your score isn't as high as you think it should be, there is a lot more to real-world influence that you can effect than is captured in these scores.

Of course, that doesn't stop thousands of third-party developers from connecting Klout scores to all sort of apps and websites. We've seen the beginnings of Klout-gating on Facebook, where visitors to a brand page are asked to sign in with Klout. They are then directed to different pages depending on their scores: wallpaper downloads for the minnows, premium complimentary experiences for the whales. Klout Perks work the same way, with brands offering free goods and services every month, but only to those with enough Klout. It's also being used to attract the 'right' crowds to 'it' events like fashion shows by using Klout-linked invitations that make sure no one with less than a Klout score of 40 gets through the door.[5]

What comes next from this? Klout jobs, where only candidates with a score above 50 get an interview? Klout real estate, where only the ones with scores in the 60s can buy property? This is a scary future. It would be easier to embrace if we could be completely certain that Klout scores were accurate. But many are not. There are those who do very little and receive high scores, while true influencers end up being undervalued because their impact can't properly be measured by today's tools.

Yet, the simplicity and relative accuracy of Klout make for a quick and easy filter – one that is being increasingly employed. I'm not saying that it's fair, but because it is easy to use it gets used. This makes Klout and other OIS an important element in developing your influence strategy.

Klout will continue to refine how they score people. They know they have competition, and they want to be accurate and trusted. So do you. Your Klout will grow if you employ exactly the sort of strategies that this book has been prescribing: find your story, share valuable social currency, find your community and share in their currency. Find community leaders, interact with them. Don't try to game it. Just do what you need to do, and watch Klout follow your growth.

Klout, Kred, Peer Index and other leaderboards are still evolving, but there is a future coming where the value we produce online with our story's social currency and our community's social capital will become part of the very fabric of the economy.

Dunbar Portfolios

As we've been building our communities, I've often mentioned to be on the lookout for influencers. It's often difficult to spot these people immediately. This is where OIS can provide a helping hand.

There are plug-ins available for your Firefox or Chrome browser that will show Klout scores and Peer Index scores immediately next to people's names in Twitter. This can give you a quick heads up, when you are monitoring the conversations in your master narrative, as to who the conversation leaders are, so you can focus your attention on them. The more conversations we have with people with high OIS, the higher our own OIS will become.[6] Remember, cocktail party etiquette prevails: wait for a chance to ask a good question or to offer some relevant insight.

We can also look at Klout topics – pages that list who is influential about certain topics. See who surfaces as the top influencers against keywords that matter to you. These are leaders in the master narrative; these are the royals we want to court.

Since we know that we are limited in active relationships to Dunbar's Number, we need to think carefully just which relationships are the most important to manage. OIS can help us decide which relationships we should be focused on.

I suggest making a list of the most influential relationships you want to cultivate – a Dunbar Portfolio. It doesn't need to be 150. In fact, start smaller so you can focus more energy on your community VIPs. Pick your Top 50 and start using all your relationship building skills.

That doesn't mean that you ignore everyone else. Don't go Hollywood. But do manage your time in a way that ensures you are building social capital with as many of the most influential members of your master narrative as you can.

You can evolve your Dunbar Portfolio over time, adding people to it as well as taking people off of it. Relationships often ebb and flow, and you need to be flexible to move with those tides. But make it a deliberate act of management. Don't let yourself be blown around the social seas like a boat without a rudder. Know where you want to go, and use OIS to help you decide who you need to be friends with to get there.

Social Capital vs Social Credit

While social capital is something that has been built up between you and other members of your community, social credit usually comes from outside your community, even from outside of your master narrative. It often comes from brands that are trying to influence you without building a relationship with you. Social capital is about small favours between friends, the sort of interactions that build lasting loyal relationships. Social credit, on the other hand, is about large gestures from strangers often designed to buy your temporary support.

When your reputation precedes you, that's social credit at work. When you get in on the VIP list at someplace you've never been because of your standing in the community, that's social credit at work. To my grandfather's generation, when the words 'your reputation precedes you' were uttered upon meeting someone for the first time, they were based on a long series of hard-earned, real-world, word-of-mouth character references. In that era, that sort of reputation would have opened doors and made it easier to begin new business relationships.

Today in the networked era, our online reputation is far more ephemeral: the sum total of activity over the past few weeks. OIS are calculated day-by-day. A strong week of posting and sharing can see your score shoot up. That would be the time to cash in.

Klout Perks are an example of fairly harmless social credit. Brands offer discounts and freebies to those who have high online influence scores. They do so knowing that reciprocity is a powerful tool of influence. Don't get me wrong. The VIP treatment is great. It's hard to say no to anything that affords you special privileges simply because you have online influence.

But accepting social credit is a risky business. OIS may still open doors just as your grandfather's reputation did, but be careful which ones you step through. Reciprocation is an old trick, and even the smallest gift can kick it off. The Hare Krishna Society would give a 'gift' of a flower, and then ask for a donation; far fewer people could say no to them after accepting the flower (which most of them didn't even want).[7] So be careful not to accept anything from those you don't like or trust – they can drag you down.

When the gifts go from small gestures (like retweets) between those fostering budding friendships to strangers offering significant gifts (like gold watches), we enter into a different game. You need to set your own ethical boundaries. Decide for yourself where you draw the line, but know that your reputation will be better the more squeaky clean you are.

As a journalist, my rules regarding accepting gifts are:

1) Nothing that can't be consumed in a single sitting. Lunches of any value are fine (as are drinks).

2) Nothing with any commercial value. Branded swag that I'm given is always re-gifted. Seldom have I kept anything. Items of any value are refused or returned to the giver.

Refusing can be difficult (nobody said ethics were easy). Sometimes I joke that, 'If it's a gift, it's too much. It it's a bribe, it's not enough.'

While it's hard to turn down nice things, there is a strong psychological reason to do so. It doesn't even require that much value. Even accepting swag has a certain inescapable affect. So beware.

Of course, since you now know how reciprocation works, you may want to turn the tables and use reciprocity to your advantage when winning over that influencer who has remained on the fringe of your community for some time. Go ahead, extend social credit. Just go easy. Those who matter the most in your community will hate the feeling that they have been bought. Bribery can turn against you (and in some cases, it can even be illegal).

What Colour is Your Whuffie?

'When reputation is wealth, only those who do good and well unto others are the richest.' This is the rationale behind 'The Whuffie Manifesto',[8] part of a growing online movement to find new ways to create and exchange value in the social media era.

Think of it this way: the term millionaire reflects quantity, not quality. The notion, which seems less implausible the more entrenched OIS become, is that we could have an economy that reflects our value in terms of the quality of our contribution to a community rather than in terms of how much money we earn or possess. The reputational value we could extract would be on a par with the amount of good we do unto others.

Whuffie is a term coined by author/activist Cory Doctorow in his novel *Down and Out in the Magic Kingdom*. In his story, whuffie replaces money and incentivizes people to do useful and creative things. Public opinion determines which things are useful and what they are worth.

> *Whuffie recaptured the true essence of money: in the old days, if you were broke but respected, you wouldn't starve; contrariwise, if you were rich and hated, no sum could buy you security and peace. By measuring the thing that money really represented – your personal capital with your friends and neighbors – you more accurately gauged your success.*[9]

Doctorow's concept of whuffie was literally a number following his characters around. As everyone had a sort of augmented reality vision, they could ping the numbers that describe the person's social standing and see them floating above them, a constant reminder to all how good or bad you were. Like our OIS, whuffie scores summed up a person at a single glance.

If all this sounds a bit too farfetched to be relevant, think again. Even without OIS, we already have a 'balance sheet' of both online and real-world conduct that is available for anyone with a web browser to review, assess and analyze.[10]

The important thing to note is that there are many colours and flavours of whuffie. Klout would suggest that Justin Bieber is the most influential singer/songwriter on earth. Some of us may disagree with that. There is too much music in the world for any one person to be the master of it all.

No one can be expert at everything. No one can be so esteemed by every community that they are leaders of them all. On the contrary, you hold influence in a narrow vertical, and mostly within a particular sub-tribe: gadget geeks are influenced by gadget geeks, fashion bloggers are influenced by fashion bloggers, bankers by bankers and so on.

So to understand our niche of influence we need to revisit our personal brand values. This isn't just about personal marketing; we need to consider our life-long reputation strategy. When we do good deeds for our community, it's not just so that they will be our virtual friends. We know that our actions, being public in social media, affect our reputation within the wider community that sees us. Your whuffie may not be fully reflected in your OIS, but the community knows if you are a valuable contributor.

Playing to the Crowd

Socialeaders need to have a certain confidence in their authority. They know that what they are doing is the right thing because they are comfortable with it. But sometimes that confidence in our own voice makes us deaf to influencers we should be connecting with.

You can think of it like a band before an audience that just wants to dance. They've got to play for the audience. If the band puts their egos first, and plays only the songs that they like to play, but that no one can dance to, then they've failed. No matter how good a musician you are, no matter how good you think the music is, the dancing was the experience that the audience wanted. They may marvel at your musical skill, but they aren't having as much fun if they can't connect with you.

We can connect better with people if we can adjust our tone to match their attitude. So we don't have to change our song – our story remains true to us – but we can adjust the key and the tempo to make it easier for others to receive. There is a lot of psychology around this, but a simple framework like DISC assessments (Dominance, Influence, Steadiness, Conscientiousness) can provide us with some insight into how we can relate better to the tone of others in social media. We don't need to run complex DISC profiling on all of our online connections in order to take advantage of this influence tool. All we need to do is to listen to how they communicate, and observe which social media channels they prefer.

Here's a quick guide to recognizing the DISC tone in others:

Dominance: These folk are all about moving things forward and getting to the point. They want the headline and the punchline. They respond well to tweets that use terminology like progress, drive, results, output and actions taken.

Influence: These are people who value connecting and motivating. They like songs in the key of fantastic, awesome and fabulous. These people are all about positive energy, post lots of motivational quotes and will partake in #FollowFriday with a near religious devotion.

Steadiness: This group talks a lot about standards and best practices, so it's best to respond to them in kind. Mention anything to do with standards, systematic approaches and methodologies. Emphasize efficiency. They likely have really well manicured LinkedIn profiles.

Conscientiousness: The group most resistant to change. They want to avoid conflict and upheaval. They want to hear about maintaining balance and productivity. They value that which is consistent, reliable and stable. Possibly best to reach these people using newsletters or other email-based networking.

Socialeaders need to be able to recognize the language of influence being used by these different groups and adapt their language, and communication channels, to suit. If you fail to connect because you can't be bothered to adjust your tone, it is your reputation and authority that will suffer the scorn of your audience.

When they trust you, the audience actually gives you authority. I have a foodie friend who always seems to know the best dish to order in every restaurant in Hong Kong. When I go for dinner with him, I don't even bother looking at the menu. I just let him order for us. That's me giving him authority.

So while we can drape ourselves in titles and trappings of authority, the real authority will come from word-of-mouth reputation in our communities – not dissimilar to how our grandfather's generation would have experienced reputation. Reputation will be a source of influence, but only so long as you continue to demonstrate good judgement, and deliver a good experience. When my foodie friend starts ordering dishes that don't taste good, I'm going to start asking for the menu.

Internet Fame May be Less Influential than We Think

If we measured social capital based on social proof, then the most influential people in the world would all be musicians. Lady Gaga, Justin Bieber and Britany Spears have been some of the most popular people in social media in recent years. This doesn't mean you should quit your job and start practising the piano.

The thing is that even really famous piano players don't really have that much influence over their fans. On AIDS Day, 1 December 2010, Alicia Keys sacrificed her digital life to help her charity Keep a Child Alive raise money for those affected by HIV/AIDS in Africa and India. She managed to get several high profile Twitter celebrities to join her, including Lady Gaga, Kim and Khloe Kardashian, Ryan Seacrest, Justin Timberlake, Elijah Wood, Serena Williams and others, in voluntarily giving up their online activities until US$1,000,000 was donated by fans.

They all shot a last Tweet and Testament YouTube video imploring fans to give to this important cause, and swearing that they would not return until the money was raised. This might have been the most poorly conceived part of the plan, as the very tool you'd want to use to call in your social capital is the one tool you can't use. Instead, this campaign then became a test for how much pure Internet fame could move people to give money. The theory goes that if the community really cares for you, then your absence will be unbearable – or at least noticed.

There was a time when I was tweeting 200 times a day. I was such a noisy part of many people's social media stream. While some complained and a few stopped following, I amassed most of my 100,000 followers during that time. Then I had to dial it back for a while – and even abruptly stopped for a few days. Immediately, I had people tweeting me and asking where I had gone. My absence was noticed. Of course, I wasn't asking anyone to pay to get me back.

But the superstars' fund raiser had asked for a million dollars before they'd return. Between them, they had at the time more than 10 million fans. They only needed to convert 1 per cent of their fan base into giving $10 each. That sounded easy enough for people who seem so popular on TV and radio. At first I predicted the whole charity event would be over before lunch. But it wasn't.

It stretched in to a second day, and then a third, and they were still nowhere near half way. Days four, five and six were no better for them (for the rest of us, life on Twitter carried on without them). By the end of the week, the celebs had grown frustrated, cut off from fans and only halfway to their million dollar goal. To end their digital fast, they had to call in pharmaceutical billionaire and philanthropist Stewart Rahr to bail them out with a cool $500,000.[11] Ouch. Maybe their fans simply didn't miss them that much. Maybe they didn't value the cause that much. Either way, it was clear that the celebs' influence outside of creating music and TV that appeals to a lot of people is limited. They have one colour of whuffie, and limited influence outside of that.

It's all about the targeted influence – relevant influencers. Don't look for the most influential person talking about a major world event, look for the person with the *most* friends talking about the event because they are actually involved in the event.

Take Away Express #14

- ☐ Influence comes in many forms. Find that which suits your style.
- ☐ Online influence scores, like Klout, are now attached to us and will affect how the web and the world interact with us.
  - The good news: mastering the ability to exchange valuable social currency among a power community with whom you hold social capital will boost your score.
- ☐ As you build social capital with enough people at a high enough level, your reputation will precede you in a more meaningful way than online influence scores reflect.
- ☐ Create action-based influence.
- ☐ Enjoy the perks of online influence scoring systems, but be wary of the risks of accepting gifts from those trying to subvert your power.

15. IT'S TRUE: SIZE DOES MATTER

'There are two types of people in social media: those who admit to wanting more followers, and those who are lying.'

– Guy Kawasaki[1]

Yes, despite my earlier advice that an influential few trumps an indifferent hoard, gaining lots and lots of followers does matter.

Most of us feel the need for popularity on some level, even if we won't always admit it. While our motivation is often driven by ego or vanity, popularity is a key part of a powerful influence tool known as 'social proof'. Social proof generates social credit for free.

But before you set a goal of becoming a twillionaire, remember that quality is still important. It's also important to understand just how many followers you need to stand out from the crowd. The answer may surprise you.

Imagine you are sitting in a quiet bar with a couple friends. Of the 10 or so people in the room, no one is even close to being a millionaire. Then Bill Gates walks into the bar. Now that he's in the room, everyone there, on average, is a millionaire. In fact, on average they've got about $500 million each. But if you looked at your own bank balance at that moment, you'd see that you are really, really below the average. But you're not alone, the vast majority, about 90 per cent of the room, are below average.

This is exactly the case with the power-law distribution in social media. We have a short head of very popular people, and a long tail of niche players. To put into perspective where you truly stand, don't just look at the average of a social network, look at what the overwhelming majority of people have.

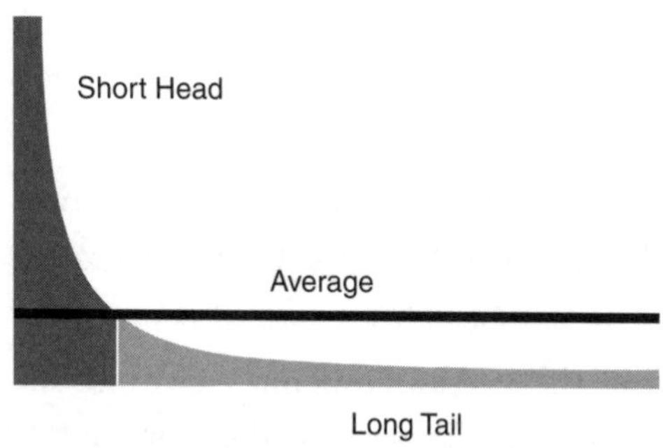

Nearly everyone is below average. You'll find fellow midfluencers around the average line.

The averages keep rising each year at a faster pace than the vast majority of people seem to be capable of keeping up with. So the trick is to merely stay as much above the average as you can, and you will find yourself in the elite minority. Keep pushing and you'll find yourself becoming part of the 1 per cent fairly quickly.

These numbers aren't hard to achieve in an authentic fashion. There is no point recruiting a thousand spam bots just so you can be above average. Falsified social proof will only work against you. And with the prevalence of ever-improving online influence scores, you can't really afford an audience that doesn't engage with you. Your fans need to be real. Those who dominate the short head tend to be celebrities offline too. I'll talk about the extreme measures you can attempt later if you are desperate to get into that space.

For now, you don't need to be the Bill Gates of your master narrative in order to utilize social proof. But you will need to make an effort to acquire a few more friends.

Social Proof

The greater the number of people who find any idea correct, the more the idea will be correct. In other words, 100,000 people can't be wrong (even if they are).

Leaderboards, like Wefollow.com, Twitterholic.com and Twittercounter.com, play to this. These are but a few of the dozens of sites designed to chart the most trafficked, most followed, most subscribed and otherwise most popular social media users.

Even though OIS are much better indicators of influence, we can't help ourselves when we see leaderboards. Everyone likes to know who has the most. Even if that doesn't really reflect any true influence, it does convey influence of social proof.

That's why we like widgets that show off how many subscribers and followers we have. We assume that the actions of so many others must indicate the truth, or some sort of insight. Our brains play tricks on us, which is why you need to consciously become much more suspicious of social proof when you see it in others. It is quite possibly faked.

Social proof can come in surprising forms. Television sitcoms have used laugh tracks for years to evoke more laughter. The studies show that hearing other people laugh, even when it's canned, by instinct is social proof of funniness – which in turn makes us laugh longer and harder at jokes that aren't all that funny.[2] We follow the crowd, even if we can only hear artificial versions of them.

People follow people often because they already have lots of followers. If you are a teenage girl, it's hard not to follow Justin Bieber if all your friends are. Social proof triggers our desire for conformity, and compliance. And social proof becomes even more powerful when it is connected to someone seen as an authority or expert.

So, to make this work for you, you first need to worry about your story and your community. All that social currency exchange and community building need to happen before you set off to establish more social proof online. No one is going to believe you really have 100,000 fans if you've only tweeted seven times.

There are other reasons for massively expanding your community, other than social proof. Larger communities mean much bigger networks, which open you up to many more opportunities. The more connections you have, the more possibilities there are to create value.

How to Make Lots More Friends

I hope you didn't skip straight to this part of the book. This is always what people want to know, hoping that there is some sort of secret formula to rapid friend growth. There isn't. There are no short cuts. Hard work only.

Those who offer you short cuts in exchange for your cash are ripping you off. Buying followers or friends will gain you nothing. Many of the operators that try to sell followers use means to add followers to your account that violate the terms of service for nearly every social media platform. Most of the followers they add to your account are spam bots. These sorts of activities can get your account suspended, and then you are out of business.

So how do we legitimately gain lots of new followers? The easiest thing you can do is follow or friend them first. Go ahead, follow lots and lots and lots of people, but be strategic. Follow lots and lots of relevant people. Look inside your community. Look at who follows each other. Follow the best of them. Set an OIS threshold and, for example, maybe only follow those with Klout scores above 40. Or if you don't want to put too much weight on OIS, set the bar lower, but still seek quality; follow only those who fulfil basic reputation requirement, real name, real picture, relevant bio and regular contributions over the last few weeks.

There are some caveats to this strategy. Many social media sites like Twitter take a dim view of aggressive churn. Churn is when you follow quickly to see who follows you back right away, then unfollow all those who don't reciprocate.

Churn happens in places where there are limits on how many people you can follow in total. So what spammers learned to exploit in the early days of social media was the ability to follow loads of people, wait a day to see which ones were drawn in by that old influence power tool reciprocation, then dump those who weren't and then follow again as many as possible the next day. They wrote black-hat software scripts that automated this process so they could sit back and relax while their machines collected thousands of followers for them. All the major social media sites now have hunter-killer software that identifies and automatically suspends those who engage in such practices.

Today on Twitter, if you follow your maximum number of people in a day, then dump all those that don't follow you back, and repeat this activity for a couple days in a row, Twitter will suspend your account. You are allowed to slowly, ever so slowly, unfollow batches of people, but don't go crazy and do this all at once. You will be exiled and all the value that you've built with your community while using that account will be lost.

Do not jeopardize all your social capital in some foolish quest to attain more social proof.

Instead, set aside some time every day to follow more of the people who fit into your narrative. Be a bit selective. Make relevant choices. The percentage that will reciprocate and follow you back will be much higher than if you just follow randomly en masse. This will allow you to continually follow more and more on Twitter, as they will increase your limits if you have a lot more people following you back.

Once you start to get an above average following, there are lots of other things that you can do that will attract followers without you having to follow them first. This is advised, as accounts that follow more than they have following them are viewed with suspicion. Even though I suggest trying to follow everyone back, it's important to have more followers than you follow. On Twitter, any ratio above 1:1

looks like you are trying too hard. The sweet spot is between .6 and .8 (i.e. following between 600–800 for every 1,000 who follow you).

Once you have established a following in the thousands, a number that easily demonstrates social proof, you can then leverage this to gain more exposure in other ways.

10 Ways to Leverage Your Social Proof

It's important to build momentum, like the snowball rolling down the hill: it keeps getting bigger the longer it keeps rolling. Here are a few ways to keep the ball rolling:

1) List Yourself on Directories

A bit like putting yourself in the Yellow Pages of old, today you can add yourself, usually for free, to dozens of social media directories like Wefollow, Twiends and Twello. Directories often let you list yourself under a few different keywords. Choose words that are relevant to you, but also be clever. Choose variants that make sure you rank as high as possible. Instead of using the word ice cream, which might get you listed in the 200th place, chose French vanilla and be in the top 10.

These directories are also leaderboards and a place where you can find lots of relevant people to follow. Have a look at who ranks about the same place as you. See if you can make some new friends – or at least keep an eye on the competition.

2) Hashtag Keywords When You Share Social Currency

On Twitter and Google+, it's possible to add the # symbol before a word. This makes it more searchable. You can likewise track those keywords to find other like-minded people who will likely follow you. Watch for others who are tagging key words. For example, if you are into ice cream, those who use tags like #doublescoop or #wafflecone would likely be people you want to check out. The same thing will happen when others watching for those hashtags see your tweets. But don't overdo it. Seeing tweet after tweet loaded with hashtag on every word is annoying.

3) Cross Post and Interconnect Everything

Every platform deserves a unique approach, but when looking for more quantity, it pays to cast a wide net. There are plenty of tools that allow content to go from blogs out across a wide array of social media sites, and then into a wide range of groups on those sites. This is the mass marketing approach to social media where you get your content out there as far as you can in hopes that as more people see it, you'll attract more followers.

4) Use Social Proof Widgets

There are plenty of little JavaScript badges that you can get for your blog. Many of these badges are provided by the services that track your metrics. You can get badges from sites like Feedburner, Klout, Twitter and Facebook. These badges provide a live running count on everything from how many subscribers you've got on your blog, how many Twitter followers you have, how many people have signed up for your newsletter to how much of your content is being shared online. Only use these once you've got numbers that you feel are respectable, otherwise it can backfire on you. Showing up at a site that only has 10 subscribers and 19 Twitter followers is hardly the most compelling social proof – it actually proves the opposite.

Widgets can also be used to automatically bring in live tweets and Facebook updates to the sidebar of your blog. If you can't show social proof, at least you can show people that you are actively sharing interesting social currency.

5) Guest Blog

Anyone with a blog knows it's hard to keep the content flowing. It can be a big help to offer to write a guest post for a blog in your industry, but something that is outside your company. Most bloggers will usually welcome a guest blog post written by an authority. In return, ensure the post has lots of links to your social media profile so you can drive that blogger's traffic to you. If readers then like what they see, you can convert them to followers.

6) Share Reports, eBooks and Presentations

It's easier than ever to create great PDF reports and small ebooks that showcase your though-leadership, data or research. The more you've been able to convert social currency into more valuable form the more you'll have to put into these reports – turn data into knowledge and wisdom. There are countless sites through which you can promote these reports.

Give your reports away as social currency, let your reports spread as far as possible. Make sure your name and a link to your social media profile are prominent so that if readers like your report, they can follow you for more insights. Also if you have any presentations, even if the slides were just put together to brief colleagues, you might want to consider putting them up on all the various slide sharing sites. Anywhere that you can get more exposure for work you've essentially already done is working smart.

7) Say Thanks

I've mentioned this several times in this book, but it warrants repeating. Saying thanks may seem too simple to work, but it has a surprisingly powerful effect. If you thank everyone whom you have an exchange with online, lots of other people notice that, it will make you more likeable, and more will start to follow.

8) Start Talking to Yourself

Showcase your expertise through something other than the written word. Host a series of webinars or podcasts, for example. There are plenty of tools online that can help you do that. Spoken word recordings are high value social currency – but also a lot more work to put together. But if you can make the time on a regular basis to make your expertise into a regular series, and if it's good, then people will subscribe and follow so that they don't miss your next episode. The fear of missing out turns casual visitors into followers. Take note that written blog posts are a lot more likely to be found through search than rich media files are. So if you create a webinar or podcast make sure to embed it in a blog post too. Also look to places where webinars and podcasts are being shared, like webex.com or iTunes. Not only are these places to help distribute your content, but also places you may find more high value community members to befriend. Follow their followers and you'll also gain more audience for your showcase.

Note that this is a challenging type of storytelling. You need to deliver a great performance without being able to see or hear the audience. Make sure that the audio quality is very good. Use good equipment (quality microphones and high-bitrate recording). New followers have a low tolerance for poor quality experiences, regardless of how good the content is.

9) How-to Videos

If audio is powerful, videos are super powerful. YouTube is one of the most used search engines in the world. We learn really well from videos, so it pays to be in this space. The best known celebrities in social media have all come from visual media. We place great social value on those whom we are able to recognize by their faces. We seldom know what award winning authors look like, but we know what award winning movie stars look like. The stars of movies and music (which is all about the videos too) are the most followed people in social media.

However, know that doing this well is harder than it looks. If you can get good at video, you will attract lots of new friends. But make sure that the production quality is excellent. Shoot in HD. Record clean sound. Make it look professional.

10) Lead an Offline Group

We talked about professional groups being part of the Big Five places to start hunting down community. Instead of just joining, why not lead one? If there are no social media societies or clubs in your neck of the woods, set one up using sites like meetup.com or even Facebook events.

Sites like meetup.com make this fairly simple. You'll still need a venue, maybe a local bar or coffee shop. The size of the meetup will determine the size of venue you need to find to host it. While there is certainly a lot more work in putting together your own meetup, meeting people in person strengthens the bonds you share online. You are less likely to misjudge your level of intimacy after you've met in real life. You'll also be more likely to gauge your social capital. As meetups grow in size, they increase your opportunity to add more and more valuable followers.

Focus on building a membership that can assist your business through referrals. In exchange, show others how to become a Socialeader.

Five More Extreme Ways to Get even More Exposure

Obscurity is the enemy, and we must do whatever it takes to break out. Anything that can gain us some more publicity or exposure to a greater audience should be considered. When all else fails, some people try throwing money at the problem. It sometimes works. So does shameless self-promotion and media whoring. These tactics aren't for everyone, but sometimes the battle for attention calls for drastic measures.

1) Offline Publicity

Try submitting an article to a magazine. You can either contribute a free article or column to a trade publication, or try to get paid by bigger consumer periodicals. Make sure there is opportunity for a bio or other means of promoting your social media handles.

If you're not up to the commitments of creating even more content yourself, let a reporter do it for you. Let bigger media outlets in your area know that you are available as an expert to comment on issues that arise through current events. This can be as simple as giving a few quotes to a reporter over the phone, or as intimidating as having a TV crew follow you around for a week. Radio is another medium that always needs guest experts to come talk on their shows. Other people's podcasts should be included in this mix as well as they have many of the same needs as radio DJs and often even more focus on your area of expertise.

If you are serious about landing mainstream media engagements, it can help to hire a public relations professional to promote you. They have the relationships that will get you the exposure you need. Make sure you use these opportunities to give the mass public your social media contact details.

2) Speaking at Conferences or Networking Events

Look for opportunities to speak at industry conferences. As a Socialeader steeped in the power of exchanging social currency, you will have a lot to offer an event organizer that most speakers don't. You can promise that you won't sell from the stage. You've got so much curated content at your disposal that it is easy to stand up and present big ideas and cool concepts. Shower the same value on the live audience that you give to your online audience. If there aren't many conferences for you to speak at, you can always try local networking events. If you aren't talking about yourself, but rather bigger issues, you may get invited back frequently to give updates. After you've provided a great presentation, it's a great time to ask people to follow you. They'll want to balance the social capital.

3) Hosting Your Own Conference

If there aren't enough conference or networking opportunities, try creating your own. If you are not satisfied creating a small meetup, go big. Hire event organizers. While they're not cheap, they can produce the show for you. All you need to do is create a quality programme: hire quality keynote speakers, invite panels and moderators, and prepare a few short speeches of your own.

This will certainly put you in the limelight, but know that it can take a few years to make a conference a success, and it's hard work.

4) Advertise on Social Media / Google / Blogs / Leaderboards

There are whole books written on how to get started buying advertising on social media sites, like *Facebook*

Advertising for Dummies. But also consider more personal approaches. Try promoting yourself on leader-boards like Twittercounter, where for a modest fee you will be placed at the top of relevant lists. Remember this is about promoting you, not your company, so the advertising budget might have to come out of your own pocket.

5) Advertise Offline

In Hong Kong, the streets are filled with gigantic neon signs for relatively small restaurants. The psychology behind the signs plays off of social proof: a really big sign indicates that the restaurant can afford such things, which means that lots of people must eat there, which means it must be good. That's not always the case, but buying advertising isn't about making your brand better; it's about attracting attention. The challenge here is how you as a brandividual can afford – or leverage if the company is paying for itself – such big advertising spends. It is saying the same thing as the neon signs: if I can afford advertising, then I must be good.

This is why celebrities who are on TV and grace the pages of magazines, or the sides of buses, have such disproportionately large followings. Remember, bigger doesn't always mean better.

Make Every Space Count

Put links to your online presence everywhere. Be creative. Wherever you can find a spot, go for it.

Business cards routinely feature people's Twitter handles or Facebook profiles or LinkedIn addresses – or sometimes all three. I've seen cards that fold out to provide two pages of links to all their social media profiles.

Email signatures are another easy place to put a 'follow me' button, or links to any place you want people to join your community.

I've even seen people wear T-shirts with QR codes on them that others can scan with their smartphones to automatically follow that person. QR codes are a type of barcode, but square and full of little dots that can be read by apps on most smartphones. The codes can be generated online to link to a specific URL, then you can print QR codes on just about anything.

These links can be included on any swag (T-shirts, hats, stickers, etc.) you want to make. But you should also include something human readable like your Twitter handle. A simple @username printed somewhere visible is all you need to get more people to follow you.

But always try to create exposure in relevant ways. They won't follow you just because they see your name. They need a reason. They need to be interested in you. This is the reality of mass marketing. There is a lot of waste.

Ultimately, none of this stuff is really essential for you to make it as a Socialeader. You may go through a phase where you want to boost your exposure, to move yourself up in the community faster. But essentially these are all campaign tactics. You have got to be ready to sustain the community with a great story week in and week out.

These extreme tactics are at the back of the book for a reason. This is not the place to start – although I've seen many brands try. This is what we do *after* we master our story, find our community and establish our online influence. It will all work much better then.

Take Away Express #15

- ☐ The vast majority of social media users are below average. Socialeaders work to stay above average.
- ☐ Social proof can be a powerful weapon of influence, but it needs to be wielded with great care. Build an authentic following on a solid foundation.
- ☐ Follow lots of relevant people every day.
- ☐ Look for opportunities to increase your exposure:
  - Connect yourself to directories and keywords.
  - Guest blog, share reports and presentations.
  - Webinars, podcasts, videos, group discussions.
  - Advertising and publicity.
  - Get in front of live crowds.
  - Get your name everywhere you can.
- ☐ This is not a starting point. Use these techniques after you have mastered your story, found your audience and established your online influence.

16. SOLVING THE UNDERPANTS PROBLEM

'An organization's ability to learn, and translate that learning into action rapidly, is the ultimate competitive advantage.'

– Jack Welch[1]

If you've not seen the cartoon *South Park*, please bear with me. In one of my favourite episodes from 1998, the town of South Park is being inundated by a big name coffee chain and the boys, Kyle, Stan, Cartman and Kenny, go on a quest to stop it from putting the local mom and pop coffee shop out of business.[2] In the process, they uncover a secret operation being run by gnomes who sneak into children's bedrooms at night and steal their underpants. Yes, gnomes and underpants. If you've not seen *South Park*, it's that kind of humour.

The boys eventually apprehend one of the gnomes and demand that he show them how business works, so that they can stop the big name coffee shop. They follow the gnome to the underground gnome headquarters where all the underpants are being collected. The boys press the gnome to explain the operation to them.

The gnome pulls down a big chart and explains:

1: Collect underpants.

2: ?

3. Profit.

The boys are left more confused than when they began.

What makes it so funny (and sad) for me is how true this chart is for so many start-up businesses. Of course, those running the start-ups can talk their way around phase 2. It gets dressed up, glossed over and buried in the hype of phases 1 and 3. But the question mark is still there.

The Underpants Problem is also the most common social media problem for beginners. They put their clever content into the social media realm, and presto, profit.

Let's back up to the 'presto' part. What exactly is supposed to happen in that middle part?

Socialeaders know that without valuable social currency, without a community with which you hold meaningful social capital, without online influence that unlocks social credit, the presto part will never happen.

Phase 2 is when you turn your influence into advantage.

Use Influence to Convert Lots of Little Guys

By little guys, I don't mean gnomes. The little guys are the many, many casual users who have yet to invest the time and energy into turning their online presence into something influential. The little guys are what the B2C Socialeaders will spend a lot of time focusing on. Currently much of that focus is on engagement.

But engagement isn't what we need to be measuring. Sure, it's useful to track how many exchanges you've had with other people online. We can track how much they view us and share us. This is an important part of building social capital, but it asks for no real commitment.

Engagement relationships are founded only on weak ties. They need to be converted into strong bonds. This conversion requires a small request, asking your community for a small commitment, which coming from a respected community authority should be a relatively easy thing to get.

What sort of conversions do you need? As big as you can get. This doesn't mean they have to click the buy button on your website (although that would count as a conversion too). But getting them to make any small commitment towards you will lead to a relationship that will spend money with you. Remember the influence power tool combo of commitment and consistency? This is what is at work here. But you'll also be putting the full host of influence power tools to work.

Conversions that leverage your authority, your social proof and your likeability, to move community members to making commitments to you, even small ones, will pay off in the long run. Psychologically we want to remain consistent to our commitments.

Socialeaders can ask fans to join 'special' mailing lists, VIP clubs or other more exclusive ways of connecting. Anything that makes them feel special. This process can be designed like a funnel with each step providing opportunities for loyal fans to express their advocacy in exchange for richer and richer experiences. Ultimately fans need ways to easily give you money in exchange for goods.

If you don't have something physical to sell, then it shouldn't take long before you can identify what it is that fans are willing to pay for access to. If it's not about goods, it's almost always about access.

Access works like this: No one wants to pay for the mp3, but raving fans will pay just about anything to get backstage passes. Access is the only true scarcity on the web. If it can't be digitally copied, then it's worth a lot more.

So what we want to measure is how much demand we are creating for access. We can measure our engagement and then measure how much conversion we are getting from that. If conversion is low, then we need to up the experience. This is about the loyalty programmes.

It's about using conversion as an opportunity for advocates to express their love, and make public commitments in exchange for access. This gets all the social capital flowing and attracts more and more of your community.

Keep giving them stuff to do. Make it fun. Make it valuable social currency. Make it part of the conversion funnel. Lather, rinse, repeat – and sustain. This means more story, more community building, more careful construction of influence, more caring, more customer service.

This will build a massive advantage over any competitor who isn't doing it.

Use Influence to Partner with Big Guys

Partnering is more for B2B than the conversion strategy above. When I say big guys, I mean other Socialeaders. I also mean people with lots of offline influence that may not be reflected online.

Imagine if the telephone was only a mass marketing tool. All we would ever receive would be telemarketers trying to sell us stuff. If you are like me, you would likely smash your phone and never buy another. The same would be true of social media. If we spam people with sales pitches, they will close their connection to us forever.

There will always be some who use social media as a mass marketing tool, but do they stop and ask, 'How many more potential customers will I lose forever?' They should forget about generating quick returns on investment and instead ask: 'How do we measure Loss on Lack of Proper Investment?'

When we invest properly in social media, we create influence which allows us to do something even more amazing. When influential business leaders, Sir Richard Branson for example, pick up the phone they don't start cold calling strangers at dinner time and pitching something at random.

Instead, they deliberately call other influential people and they make very targeted pitches that provide mutual opportunity.

Socialeaders don't use social media to mass market like a telemarketer. They use social media to cultivate the influence and relationships that allow them to identify and seize bigger opportunities.

So solving the Underpants Problem doesn't have to be such a mystery: it's about connecting the right people with value in the right way. Through service, understanding and relationships, social media bring opportunities to your business and to you professionally.

Future-Proof Yourself

Solving the Underpants Problem is a long-term effort. Sustaining all this activity over the long haul is one of the greatest challenges we face as Socialeaders. The term social media fatigue is used a lot these days, and will be with us for some time to come. Here are some tactics to help keep you in the game:

1) Stay Connected

It's hard to stay enthused. Social media fatigue hits everyone at some point. It's really just social fatigue. Managing so many relationships while keeping up on all the latest and greatest social currency will take its toll. Remember the race is long; you must pace yourself. Look to the audience for encouragement, draw from their passion. Stick with it, and it will get easier as you find a stride that you can maintain.

2) Adapt to Change

Don't get too locked in to any one social media platform. I've tried to write this book with as much social media agnosticism as possible because I know that the sands shift frequently and unexpectedly. But the bedrock remains. Stand true to your values but expect that you'll have to find new ways to express them.

3) Move with the Tribe

The community is everything. If everyone decides to start hanging out on a new social media platform, then you need to go with them. That doesn't mean you need to decamp from the old hangout though. You'll find that it is usually just the early adopters who move at first. Many of those may be your most important community influencers, so you'll need to go with them. But stay connected to the laggards too.

4) Experiment and Measure Results

Churning out the same sort of social currency for the same groups of people all the time will produce a diminishing return. You need to change it up. Try new things. But always measure the results so that you can see what has positive and negative effects.

5) Learn from Mistakes

It's best when we can learn from the mistakes of others. But sometimes we make mistakes too. That's fine, as long as we react quickly to rectify the mistake and take steps to prevent a repeat performance.

6) Lead by Example

Just because all the other kids are doing, doesn't mean you do to. At least that's what my mom used to say. She was probably right. Be a Socialeader, not Sociafollower. This requires character and values. Principles earn respect and loyalty.

7) Be Ready to Seize New Opportunities

The coolest things that happen in social media are serendipitous. They explode spontaneously and rain fortune down on those who were quick enough to spot the opportunity.

Take Away Express #16

- ☐ Don't get stuck on the Underpants Problem. Move beyond posting content and providing service. Once

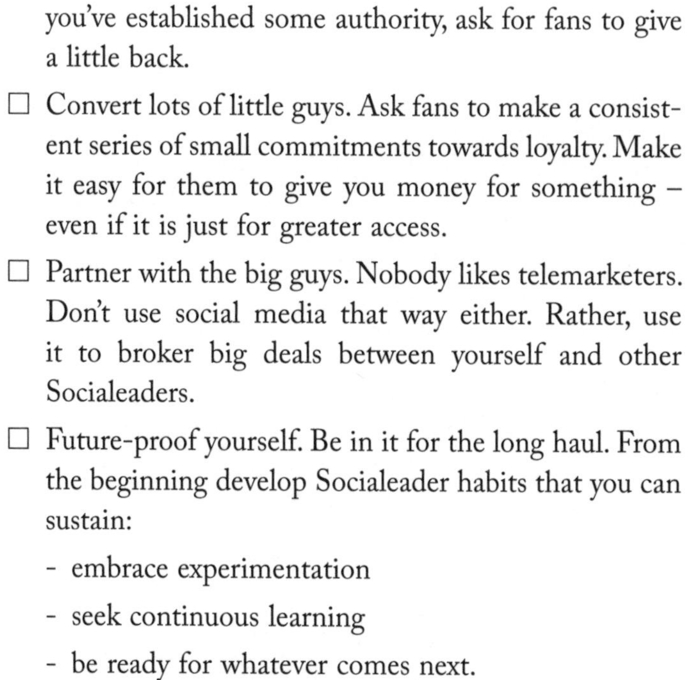

you've established some authority, ask for fans to give a little back.

- ☐ Convert lots of little guys. Ask fans to make a consistent series of small commitments towards loyalty. Make it easy for them to give you money for something – even if it is just for greater access.
- ☐ Partner with the big guys. Nobody likes telemarketers. Don't use social media that way either. Rather, use it to broker big deals between yourself and other Socialeaders.
- ☐ Future-proof yourself. Be in it for the long haul. From the beginning develop Socialeader habits that you can sustain:
  - embrace experimentation
  - seek continuous learning
  - be ready for whatever comes next.

17. ALL ABOARD THE SOCIAL TRAIN

'A powerful global conversation has begun. Through the Internet, people are discovering and inventing new ways to share relevant knowledge with blinding speed. As a direct result, markets are getting smarter – and getting smarter faster than most companies.'

– *The Cluetrain Manifesto*[1]

Begin with the end in mind. A massive organizational shift is coming – to the way we work and the way we live. It's everywhere. There is no escaping it. Everyone is a media company now. And, as the effects of online influence remap the world, you will want to ensure a seat at the table for yourself and for your family. Your best chance is to become a Socialeader and rise with the tide.

The divide between the social 'haves' and 'have nots' will be very real. Those who fail to chase this opportunity today are placing their future in jeopardy. We have seen the beginnings of a reputation economy take hold, with velvet ropes going up around goods and services reserved for only those with enough online influence.

That online influence doesn't just happen. It can't truly be bought – no matter how many digital agencies say they can sell it to you. Socialeaders must find their own stories. You must be clear in your heart what you stand for. And you must step confidently onto the public stage and start meeting people.

To be a social 'have', seek out the connections you need to get to where you want to go. Foster those connections through the exchange of social currency, letting your curation tell your story, rather than coming on too strong with a sales pitch. People want to be wooed. But remember, loyalty can't be bought. It is earned through actions.

Actions speak louder than words, which means that we need to put others before ourselves. Don't let your quest for influence become an ego project. Instead, look out for those around you. Build a reputation as someone who takes care of his or her community. It's how you help others that is going to make the real distinction in your reputation.

Be accountable to your audiences, for it is they who give you your authority. They are the ones who will back you in a fight. Don't just 'build' community, become a part of one. Be bold and take a leadership role if you can. Get involved with the work of other people; let that community participate in your work. This is the beginning of an era of collaboration, and those of us who have come from a previous era need to make a conscious decision to open up in a way that is much more natural to the digital natives.

But the digital natives also need role models. We need to set the example for the professional way to conduct ourselves through social media – it's no different from the phone, which can also be misused at work or used to close a big deal. At the time of writing this book, only 19 per cent of *Fortune* 100 CEOs were on Facebook and only 13 were on LinkedIn.[2] Today's leaders must do more to demonstrate to the next generation of leaders how it is done.

You need to look to building and leveraging long-term partnerships, instead of burning bridges with spammy promotions. We are at the beginnings of cultivating an online army that will be with us for the rest of our lives.

The rest of our lives. Say that out loud and let it sink in. We have decades of social media to come in our lives. Start connecting the dots between where you are today and where you want to be in just a few years. You are not going to get there solely offline, at least not for much longer.

Don't leave your future to chance. Don't be the boat without the rudder. Be proactive in building your influence: manage your Dunbar Portfolio; connect to the right people, not the most people; claim a larger mind share of the master narrative. Be confident in yourself – you can be a significant online player. This isn't the mass media era where the limelight was reserved for the anointed few. Today, all the online influence you want is there for the taking. Go get it.

Online influence, once you've had a taste of it, will be hard to let go of. But don't let it change you. Keep it real. Sure, go ahead and enjoy some of the spoils of online influence, but remember that any social credit you accept will force you to find a way to balance the social capital.

Don't let your reputation be tarnished by the acts of others. Guard it jealously, for it is your future – and everything that happens online stays online. Search engines don't forget. Begin today with that in mind.

The Socialeader train is now leaving the station. If you run, you can still catch it.

REFERENCES

Chapter 1 – Why We Must Share

1. Douglas Rushkoff (2000), 'Second Sight', http://www.guardian.co.uk/technology/2000/jun/29/onlinesupplement13, *The Guardian*, 29 June.
2. R.I.M. Dunbar (2004), 'Gossip in Evolutionary Perspective', *Review of General Psychology*, Vol. 8, No. 2, 100–110.
3. Ibid.
4. Ibid.
5. Jamie Notter and Maddie Grant (2011), *Humanize: How People-centric Organizations Succeed in a Social World*, Indianapolis, Que Publishing Limited.
6. Brian Solis (2010), 'Social "Me"dia and the Evolving Twitter Egosystem', www.briansolice.com.

Chapter 2 – A Better Story of You

1. Robert Fulford (1999), *The Triumph of Narrative*, New York, Broadway Books.

2. Jonah Berger (2011), 'Arousal Increases Social Transmission of Information', *Psychological Science*, July 2011.
3. Charles Arthur (2008), 'Why Rickrolling is Bad for You', *The Guardian*, http://www.guardian.co.uk/technology/2008/dec/05/youtube-rickroll-link-economy, 5 December.
4. Dan Robles (2010), 'Monetizing Social Currency', http://www.relationship-economy.com/?p=11551, The Relationship Economy, 14 August.
5. Steven Levy (2011), 'Inside Google Plus', *Wired*, October.

Chapter 4 – On with the Show

1. Saul Hansell (2008), 'Zuckerberg's Law of Information Sharing', http://bits.blogs.nytimes.com/2008/11/06/zuckerbergs-law-of-information-sharing/ *New York Times*, 25 October.
2. Facebook Press Page: http://www.facebook.com/press/info.php?statistics.
3. Clay Shirky (2008), *Here Comes Everybody*, London, Penguin Books.
4. Tom Foremski (2010), http://www.everycompanyisamediacompany.com/.
5. Tom Peters (1997), 'The Brand Called You', *Fast Company*, September.
6. Bill Cosby (2009), 'Bill Cosby's Keynote Speech', http://www.youtube.com/watch?v=BY-WFfajWq8.

Chapter 5 – Whose Story Is It?

1. Elaine Partnow (1978), Quotable Women, Anchor Books.
2. Deloitte LLP (2009) *Ethics & Workplace Survey* http://www.deloitte.com/assets/Dcom-UnitedStates/Local%20Assets/Documents/us_2009_ethics_workplace_survey_220509.pdf.

3. Janet Lowe (2007), *Waren Buffet Speaks*, Hoboken, John Wiley & Sons Inc..
4. David Armano (2009), 'The Age of Brandividualsim', http://experiencematters.criticalmass.com/2009/01/23/the-age-of-brandividualism/ Experience Matters, 23 January.
5. Mia Dand (2011) 'Hewlett-Packard: Building a Global Social Media Center of Excellence', http://vimeo.com/23072990.
6. Mike Sweeny (2011) 'Five Ways to Get Your Non-marketing Employees to Create Content', http://socialmediatoday.com/mike-sweeney/294506/5-ways-get-your-non-marketing-employees-create-content Social Media Today, 11 May.
7. 2011 Edelman Trust Barometer Key Findings Presntation, http://www.edelman.com/trust/2011/uploads/Edelman%20Trust%20Barometer%20Global%20Deck.pdf.

Chapter 6 – How to Never Run out of Interesting Things to Talk About

1. Patton Oswalt (2010), 'Wake Up, Geek Culture. Time to Die', *Wired*, December.
2. MG Siegler (2010), 'Every 2 Days We Create As Much Information As We Did up to 2003', http://techcrunch.com/2010/08/04/schmidt-data/ Techcrunch, 4 August.
3. Nick Bilton (2011), 'Mainstream Media Still Drive Majority of Twitter Trends', http://bits.blogs.nytimes.com/2011/02/15/mainstream-media-still-drives-majority-of-twitter-trends/ *New York Times*, 15 February.
4. Timothey Bickmore (1999), 'A Computational Model of Small Talk', http://web.media.mit.edu/~bickmore/Mas962b/ May.

Chapter 7 – Don't Just Tell. Perform!

1. Dan Zarrella (2010), 'When's the Best Time to Publish Blog Posts', http://www.problogger.net/archives/2010/12/06/

whens-the-best-time-to-publish-blog-posts/ Problogger, 6 December.

2. Guy Kawasaki (2009), 'The Art of the Repeat Tweet', http://holykaw.alltop.com/the-art-of-the-repeat-tweet Holy Kaw! 11 October.
3. Ok Go (2009), 'A Million Ways', EMI – http://www.youtube.com/watch?v=M1_CLW-NNwc YouTube, 26 February.
4. Craigjay (2006), 'OK Go Phenomenon', http://www.youtube.com/watch?v=kddQD2NIHBw YouTube, 3 August.
5. Ok Go (2009) 'Here It Goes Again', EMI – http://www.youtube.com/watch?v=dTAAsCNK7RA YouTube, 26 February.
6. Ok Go (2010) 'This Too Shall Pass – Rube Goldberg Machine version', Paracadute Recordings http://www.youtube.com/watch?v=qybUFnY7Y8w YouTube, 1 March.
7. Dylan Tweney (2010), 'How Ok Go's Amazing Rube Goldberg Machine Was Built', http://www.wired.com/gadgetlab/2010/03/ok-go-rube-goldberg/ Wired Gadget Lab, 2 March.
8. OK Go (2010), 'White Knuckles' Paracadute Recordings, http://www.youtube.com/watch?v=nHlJODYBLKs YouTube, 19 September.
9. Priya Elan (2010), 'Here They Go Again: new OK Go video is a White Knuckle ride', http://www.guardian.co.uk/music/2010/sep/25/ok-go-white-knuckles *The Guardian*, 25 September.
10. Chris Anderson (2011) 'Film School', *Wired*, January.

Chapter 8 – Finding Your Community

1. Brian Solis (2010), 'A Conversation about You, Social Currency and Social Capital', http://www.briansolis.com/2010/12/a-conversation-about-you-social-currency-and-social-capital/ 22 December.
2. Victoria Harres (2009), 'Why People Twitter – in one word', http://socialmediatoday.com/SMC/137567 28 October.

3. R.I.M. Dunbar (2004), 'Gossip in Evolutionary Perspective', *Review of General Psychology*, Vol 8, No 2, 100–110.

Chapter 9 – Make 'Your' Story Feel More Like Service

1. John Dewey (1916), *Democracy and Education*, New York, MacMillan.
2. Jeremiah Owyang (2011), 'Social Media Crises on Rise: Be Prepared by Climbing the Social Business Hierarchy of Needs', http://www.web-strategist.com/blog/2011/08/31/report-social-media-crises-on-rise-be-prepared-by-climbing-the-social-business-hierarchy-of-needs/ 31 August.
3. Ron Ploof (2008), 'Ford, Fansites and Firefighting', http://ronamok.com/2008/12/17/ford-fansites-and-firefighting/ 17 December.
4. Rebecca Reisner (2009), 'Comcast's Twitter Man', http://www.businessweek.com/managing/content/jan2009/ca20090113_373506.htm Bloomberg Businessweek, 13 January.
5. SocialCRM 2011, http://www.oursocialtimes.com/socialcrmnewyork/.

Chapter 10 – Friend Relationship Management

1. George Orwell (1946), *Animal Farm*, New York, Harcourt and Brace.
2. Michael Heilemann (2010), 'Lessons from the Chewbacca Incident', http://binarybonsai.com/2010/09/27/chewie-stats/ The Binary Bonsai, 27 September.
3. Seth Godin (2009), 'Dunbar's Number isn't just a number, it's the law', http://sethgodin.typepad.com/seths_blog/2009/10/the-penalty-for-violating-dunbars-law.html Seth Godin's Blog, 26 October.
4. The Times 100, 'Red Bull: Engaging consumers through word of mouth marketing', http://www.thetimes100.co.uk/

case-study–engaging-consumers-through-word-of-mouth-marketing–172-444-1.php.
5. Niall Harbison (2011), 'Red Bull's smart use of social media and branded content', http://thenextweb.com/socialmedia/2011/08/27/red-bulls-smart-use-of-social-media-and-branded-content/ The Next Web, 27 August.

Chapter 11 – Friends in Need

1. John Robert Colombo (1980), *Popcorn in Paradise: The Wit and Wisdom of Hollywood*, Holt, Rinehart and Winston.
2. Barry Silverstein (2011), 'Loyalty Has Its Rewards – If You Claim Them', http://www.brandchannel.com/home/post/2011/04/22/Loyalty-Has-Its-Rewards-Or-Does-It.aspx BrandChannel, 22 April.
3. Vivaldi Partners (2010), '$ocial Currency Study | US 2010', https://docs.google.com/viewer?url=http%3A%2F%2Fimages.fastcompany.com%2FVivald-iPartners_Social-Currency.pdf.
4. Mónica Guzmán (2011), 'Beware Personal Armies', http://www.geekwire.com/2011/beware-personal-armies 18 October.
5. Rachel Quigley (2011), 'Hunt the Douche', http://www.dailymail. co.uk/news/article-2048367/Bimbos-waitress-Victoria-Liss-launches-hunt-tipper-Andrew-Meyer.html *Mail Online*, 13 October.

Chapter 12 – Don't Be A Social Media Snob

1. John Buchan (1940), *Pilgrim's Way*, Houghton Mifflin Company.
2. Kevin Kelly (2008), '1,000 True Fans', http://www.kk.org/thetechnium/archives/2008/03/1000_true_fans.php, The Technium, 4 March.

Chapter 13 – Take it Outside

1. J.R.R. Tolkien (1954), *The Fellowship of the Ring*, London, George Allen & Unwin.
2. Malcolm Gladwell (2010), 'Small Change: Why the revolution will not be tweeted', *New Yorker,* 4 October http://www.newyorker.com/reporting/2010/10/04/101004fa_fact_gladwell?currentPage=all.
3. Scott Beale (2008), 'Organizing Spontaneous Parities at SXSW via Twitter', http://laughingsquid.com/organizing-spontaneous-parties-at-sxsw-via-twitter/, Laughing Squid, 9 March.
4. Denise Ryan (2011), 'Thousands stream into Vancouver to clean up after riot', *The Vancouver Sun*, 16 June.

Chapter 14 – What Is Influence?

1. Geoff Livingston (2011), 'The State of Influencer Theory Infographic', http://geofflivingston.com/2011/07/15/the-state-of-influencer-theory-infographic/, GeoffLivingston.com, 15 July.
2. Robert Cialdini (1984), *Influence: The Psychology of Persuasion*, New York, HarperCollins.
3. Edelman Trust Barometer Key Findings (2011) http://www.edelman.com/trust/2011/.
4. Robert Cialdini (1984), *Influence: The Psychology of Persuasion*, New York, HarperCollins.
5. Sherilynn Macale (2011), 'You must have a Klout score of 40 or more to get into this Fashion's Night Out party', http://thenextweb.com/socialmedia/2011/09/10/you-must-have-a-klout-score-of-40-or-more-to-get-into-this-fashions-night-out-party/, The Next Web, 10 September.
6. Ash Rust (2011), 'A More Accurate, Transparent Klout Score', http://corp.klout.com/blog/2011/10/a-more-accurate-transparent-klout-score/, Klout Corporate Blog, 26 October.

7. Robert Cialdini (1984), *Influence: The Psychology of Persuasion*, New York, HarperCollins.
8. 'The Whuffie Manifesto', http://thewhuffiebank.org/static/manifesto, The Whuffie Bank, 20 January.
9. Doctorow, Cory (2003), *Down and Out In The Magic Kingdom*, New York, Tor Books.
10. Tara Hunt (2009), *The Whuffie Factor*, New York, Crown Business.
11. Emily Smith (2010), 'Frustrated celebs get back on Twitter thanks to donation from Stewart Rahr', http://www.nypost.com/p/pagesix/frustrated_celebs_get_back_stewart_I2xWuZNQF3XLXVWMXcRxTM, *New York Post*, 7 December.

Chapter 15 – It's True: Size Does Matter

1. Guy Kawasaki (2011), http://twittercounter.com/pages/premium, Twittercounter.com.
2. Robert Cialdini (1984), *Influence: The Psychology of Persuasion*, New York, HarperCollins.

Chapter 16 – Solving the Underpants Problem

1. Robert Slater (1999), *Jack Welch and the GE Way*, New York, McGraw-Hill.
2. Trey Parker and Matt Stone (1998), 'Gnomes', http://www.southparkstudios.com/full-episodes/s02e17-gnomes, South Park Studios, 16 December.

Chapter 17 – All about the Social Train

1. Rick Levine, Christopher Locke, Doc Searls and David Weinberger (2000), *The Cluetrain Manifesto*, New York, Perseus Books.
2. Allan Biggar (2011) 'Social Media for CEOs', http://www.independent.co.uk/news/business/sme/social-media-for-ceos-2375695.html, *The Independent*, 28 October.

ABOUT JAY OATWAY

Celebrated as a leading social media authority in Asia-Pacific, **Jay Oatway** is a popular public speaker, delivering seminars and keynotes to business leaders, as well as providing expert testimonials for the media. *Marketing* magazine calls Jay Oatway 'Hong Kong's answer to Twitter royalty'.

Jay has built an influential worldwide following of more than 100,000 Twitter users, through sharing in technology-related social currency.

Jay was born and raised in Canada and moved to Hong Kong in 1997 to cover the Handover as a journalist. He fell in love with the fast-paced, highly networked, always-on lifestyle of the city and made it his home.

As a former journalist who has tapped into the latest ways to share stories and build social capital with a community, he has developed a unique first-hand experience as to how the social media landscape is taking shape. From his blogs to his Facebook fan page to his iPhone/Android apps and beyond, Jay has a pervasive online presence, which continues to grow.

Jay also provides social media training on how to tell stories that captivate, how to grow a community that matters and how to cultivate online influence that can accelerate careers in the social media age.

INDEX